HOW PLEASURE WORKS

Paul Bloom is a Professor of Psychology at Yale University. His research explores how children and adults understand the physical and social world, with special focus on morality, religion, fiction and art. He has won numerous awards for his research and teaching. He is past-president of the Society for Philosophy and Psychology, and co-editor of *Behavioral and Brain Sciences*. Bloom has written for scientific journals such as *Nature and Science*, and for newspapers such as the *New York Times*, the *Guardian* and the *Atlantic*. He is the author or editor of four books, including *How Children Learn the Meaning of Words* and, most recently, *Descartes' Baby: How the Science of Child Development Explains What Makes Us Human*.

ALSO BY PAUL BLOOM

How Children Learn the Meanings of Words

Descartes' Baby: How the Science of Child Development Explains What Makes Us Human

Language Acquisition: Core Readings (editor)

Language and Space (co-editor)

Language, Logic, and Concepts (co-editor)

PAUL BLOOM

How Pleasure Works

Why We Like What We Like

VINTAGE BOOKS
London

Published by Vintage 2011

6 8 10 9 7 5

First published in Great Britain in 2010 by The Bodley Head

Vintage
Random House, 20 Vauxhall Bridge Road,
London SW1V 2SA

www.vintage-books.co.uk

Addresses for companies within The Random House Group Limited can be found at: www.randomhouse.co.uk/offices.htm

The Random House Group Limited Reg. No. 954009

A CIP catalogue record for this book is available from the British Library

ISBN 9780099548768

Printed and bound by CPI Group (UK) Ltd, Croydon, CR0 4YY

For my father, Bernie Bloom

CONTENTS

Preface xi

1. The Essence of Pleasure 1

2. Foodies 25

3. Bedtricks 55

4. Irreplaceable 91

5. Performance 117

6. Imagination 155

7. Safety and Pain 177

8. Why Pleasure Matters 203

Notes 223
References 241
Index 265

PREFACE

THERE IS AN ANIMAL ASPECT TO HUMAN PLEASURE. WHEN I come back from a run with my dog, I collapse onto the sofa, she onto her dog bed. I drink a glass of cold water, she laps from her bowl, and we're both a lot happier.

This book is about more mysterious pleasures. Some teenage girls enjoy cutting themselves with razors; some men pay good money to be spanked by prostitutes. The average American spends over four hours a day watching television. The thought of sex with a virgin is intensely arousing to many men. Abstract art can sell for millions of dollars. Young children enjoy playing with imaginary friends and can be comforted by security blankets. People slow their cars to look at gory accidents, and go to movies that make them cry.

Some of the pleasures that I will discuss are uniquely human, such as art, music, fiction, masochism, and religion. Others, such as food and sex, are not, but I will argue that the pleasure that humans get from these activities is substantially different from that of other creatures.

The main argument here is that pleasure is deep. What matters most is not the world as it appears to our senses. Rather, the enjoyment we get from something derives from what we think that thing is. This is true for intellectual pleasures, such as the appreciation of paintings and stories, and also for pleasures that seem simpler, such as the satisfaction of hunger and lust. For a painting, it matters who the artist was; for a story, it matters whether it is truth or fiction; for a steak, we care about what sort of animal it came from; for sex, we are strongly affected by who we think our sexual partner really is.

This theory of pleasure is an extension of one of the most interesting ideas in the cognitive sciences, which is that people naturally assume that things in the world—including other people—have invisible essences that make them what they are. Experimental psychologists have argued that this essentialist perspective underlies our understanding of the physical and social worlds, and developmental and cross-cultural psychologists have proposed that it is instinctive and universal. We are natural-born essentialists.

In the first chapter, I introduce the theory of essentialism and argue that it can help explain the mysterious pleasures of everyday life. The next six chapters explore different domains. Chapters 2 and 3 look at food and sex. Chapter 4 is about our attachments to certain everyday objects, including celebrity memorabilia and security blankets. Chapter 5 is about art and other performances. Chapters 6 and 7 are about the pleasures of the imagination. Each of these chapters can be read independently. The final chapter explores some broader implications, and ends with some speculations about the appeal of science and religion.

The plan throughout this book is to understand the nature of pleasure by looking at its developmental origin in individuals and its evolutionary origin in our species. The study of origins is a useful source of insight. As the biologist D'arcy Thompson famously put it, "Everything is the way it is because it got that way." Still, the very mention of evolution in the context of psychology tends to raise both red flags and red herrings, so some clarification might help.

For one thing, *evolutionary* does not mean "adaptationist." Many significant aspects of human psychology are adaptations—they exist because of the reproductive advantages that they gave to our ancestors—and I discuss some of these throughout this book. But other aspects of the mind are by-products; they are, to use a term introduced by the evolutionary biologists Stephen Jay Gould and Richard Lewontin, *spandrels*. This is particularly the case for pleasure. Many people enjoy pornography, for instance, but there is no reproductive advantage associated with spending one's days and nights looking at pictures and videos of attractive naked people. The appeal of pornography is an accident: a by-product of an evolved interest in actual naked people. Similarly, the story of the depth of pleasure is, I think, mostly the story of an accident. We have evolved essentialism to help us make sense of the world, but now that we have it, it pushes our desires in directions that have nothing to do with survival and reproduction.

Evolved also does not mean "stupid" or "simple." I recently talked about the pleasures of fiction at a seminar in an English department, and one of the participants told me afterward that he was surprised by my approach. He said that it wasn't as awful as he had thought it would be. He had expected me to present some simpleminded reductionist biological story and was pleased that I spoke instead about the intense interest people have in the mental states of the author, and about the rich and complex intuitions that underlie our enjoyment of stories.

It was nice to make an English professor happy, but embarrassing too. I thought I *was* presenting a simpleminded reductionist biological story. His comment made me realize that I am defending two claims that don't usually go together: first, that everyday pleasure is deep and transcendent, and, second, that everyday pleasure reflects our evolved human nature. These might seem to clash. If pleasure is deep, you might reason, it must be cultural and learned. If pleasure has evolved, then it should be simple; we should be wired to respond in certain ways to certain stimuli, in a way that is perceptual, low-level, and superficial—that is, stupid.

So I am aware that the claims made in this book—that pleasure draws upon deep intuitions, that it is smart, *and* that it is evolved and universal and largely inborn—are unusual. Still, I hope to convince you that they are true. I am also going to argue that they really matter. There are serious gaps in the modern science of the mind. The psychologist Paul Rozin points out that if you look through a psychology textbook, you will find little or nothing about sports, art, music, drama, literature, play, and religion. These are central to what makes us human, and we won't understand any of them until we understand pleasure.

EVERYONE HAS something interesting to say about pleasure, and many of the ideas here grew out of discussions with family, friends, students, colleagues, and the occasional stranger on a plane. But I want to mention the influence of seven scholars who have thought deeply about these issues: Denis Dutton, Susan Gelman, Tamar

Gendler, Bruce Hood, Geoffrey Miller, Steven Pinker, and especially Paul Rozin. I disagree with each of them in certain regards, but much of this book is a response to their ideas, and I'm glad to be able to acknowledge this intellectual debt.

I am very grateful to my agent, Katinka Matson. Very early in the process, she helped me realize what I wanted to say in this book, and she was later highly supportive when I needed advice or was having an anxiety attack. I also thank my editor at Norton, Angela von der Lippe, for her faith in this project, her wise counsel throughout, and her superb comments on an earlier version of this manuscript. And I am grateful to Carol Rose for her precise and artful copyediting.

There couldn't be a better community of scholars than the Yale psychology department, and I thank my colleagues, and particularly my graduate students and postdoctoral fellows, for their support and patience as I wrote this book. Marcia Johnson was department chair during this period and deserves a lot of credit for nurturing this supportive and stimulating intellectual environment.

Some of this book describes experiments that I did in collaboration with other scholars, including Melissa Allen, Michelle Castaneda, Gil Diesendruck, Katherine Donnelly, Louisa Egan, Susan Gelman, Joshua Goodstein, Kiley Hamlin, Bruce Hood, Izzat Jarudi, Ute Leonards, Lori Markson, George Newman, Laurie Santos, David Sobel, Deena Skolnick Weisberg, and Karen Wynn. I thank all of them.

I also thank those who were kind enough to make suggestions, answer questions, or read specific passages: Woo-kyoung Ahn, Mahzarin Banaji, Benny Beit-Hallahmi, Walter Bilderback, Kelly Brownell, Emma Buchtel, Susan Carey, Emma Cohen, Lisa DeBruine, Rachel Denison, Denis Dutton, Brian Earp, Ray Fair,

Deborah Fried, Susan Gelman, Daniel Gilbert, Jonathan Gilmore, Peter Gray, Melanie Green, Lily Guillot, Colin Jager, Frank Keil, Marcel Kinsbourne, Katherine Kinzler, Daniel Levin, Daniel Levitin, Ryan McKay, Geoffrey Miller, Kristina Olson, Karthik Panchanathan, David Pizarro, Murray Reiser, Laurie Santos, Sally Satel, Michael Schultz, Mark Sheskin, Marjorie Taylor, Ellen Winner, Charles Wysocki, and Lisa Zunshine. I thank the participants in my seminar on the cognitive science of pleasure for an engaging semester of discussion and debate. And I am particularly grateful to those brave souls who provided extensive comments on an earlier draft of this book: Bruce Hood, Gregory Murphy, Paul Rozin, Erica Stern, Angela von der Lippe, and Deena Skolnick Weisberg. I'm sure that I will regret not taking all of their advice.

My family—in Connecticut, Massachusetts, Ontario, Quebec and Saskatchewan—has been a continued source of support. My sons, Max and Zachary, are now too old to provide interesting developmental data, but the plus side is that they have become smart, insightful, and funny conversationalists, and I have benefited from my many discussions with them about the ideas in this book. My greatest debt, as usual, is to my collaborator, colleague, and wife, Karen Wynn. I thank her for all of the ideas, advice, support, and, most of all, pleasure.

HOW PLEASURE WORKS

1

THE ESSENCE OF PLEASURE

HERMANN GOERING, THE DESIGNATED SUCCESSOR TO ADOLF Hitler, was waiting to be executed for crimes against humanity when he learned about the pleasure that had been stolen from him. At that moment, according to one observer, Goering looked "as if for the first time he ha[d] discovered there was evil in the world."

This evil was perpetrated by the Dutch painter and art collector Han van Meegeren. During World War II, Goering gave 137 paintings, with a total value of what would now be around $10 million, to van Meegeren. What he got in return was *Christ with the Woman Taken in Adultery*, by Johannes Vermeer. Like his boss, Goering was an obsessive art collector and had already plundered much of Europe. But he was a huge fan of Vermeer, and this was the acquisition that he was most proud of.

After the war ended, Allied forces found the painting and learned whom he had gotten it from. Van Meegeren was arrested and charged with the crime of selling this great Dutch masterpiece to a Nazi. This was treason, punishable by death.

After six weeks in prison, van Meegeren confessed—but to a different crime. He had sold Goering a fake, he said. It was not a Vermeer. He had painted it himself. Van Meegeren said that he had also painted other works thought to be by Vermeer, including *The Supper at Emmaus*, one of the most famous paintings in Holland.

At first, nobody believed him. To prove his case, he was asked to produce another "Vermeer." Over the span of six weeks, van Meegeren—surrounded by reporters, photographers, and television crews, and high on alcohol and morphine (the only way he could work)—did just that. As one Dutch tabloid put it: "HE PAINTS FOR HIS LIFE!" The result was a Vermeer-like creation that he called *The Young Christ Teaching in the Temple*, a painting that was obviously superior to the one he had sold to Goering. Van Meegeren was found guilty of the lesser crime of obtaining money by deception and sentenced to a year in prison. He died before serving his sentence and was thought of as a folk hero—the man who had swindled the Nazis.

We are going to return to van Meegeren later in the book, but think now about poor Goering and how he must have felt when he was told that his painting was a forgery. Goering was an unusual man in many ways—almost comically self-obsessed, savagely indifferent to the suffering of others; he was described by one of his interviewers as an amiable psychopath—but there was nothing odd about his shock. You would have felt the same. Part of this is the humiliation of being duped. But even if there had been no betrayal at all, but an innocent mistake, still, the discovery would strip away a certain pleasure. When you buy a painting that is thought to be a Vermeer, part of the joy that it gives is based on the belief about who painted it. If this belief turns out to be wrong, that pleasure will fade. (Conversely—and such cases have occurred—if you discover that a painting you had thought was a copy or imitation is actually an original, it will give more pleasure and its value will increase.)

It is not just art. The pleasure we get from all sorts of everyday objects is related to our beliefs about their histories. Think about the following items:

- a tape measure that was owned by John F. Kennedy (sold in auction for $48,875);
- the shoes thrown at George W. Bush by an Iraqi journalist in 2008 (for which a Saudi millionaire reportedly offered $10 million);
- another thrown object, the seventieth home run baseball hit by Mark McGwire (bought by Canadian entrepreneur Todd McFarlane, who owns one of the finest collections of famous baseballs, for $3 million);
- the autograph of Neil Armstrong, the first man on the Moon;
- swatches of Princess Diana's wedding dress;
- your baby's first shoes;
- your wedding ring;
- a child's teddy bear.

These all have value above and beyond their practical utility. Not everyone is a collector, but everyone I know owns at least one object that is special because of its history, either through its relation to admired people or significant events or its connection to someone of personal significance. This history is invisible and intangible, and in most cases there is no test that can ever distinguish the special object from one that looks the same. But still, it gives us pleasure and the duplicate would leave us cold. This is the sort of mystery that this book is about.

ANIMAL PLEASURES, HUMAN PLEASURES

Some pleasures are easier to explain than others. Consider the question of why we like to drink water. Why is there so much joy in quenching thirst, and why is it torture to deprive someone of water for a long period? Well, that is an easy one. Animals need water to survive, and so they are motivated to seek it out. Pleasure is the reward for getting it; pain is the punishment for doing without.

This answer is both simple and correct, but it raises another question: Why do things work out so nicely? It is awfully convenient that, to mangle the Rolling Stones lyric, we can't always get what we want—but we want what we need. Of course, nobody thinks that it is a lucky accident. A theist would argue that this connection between pleasure and survival is established through divine intervention: God wanted His creatures to live long enough to go forth and multiply, so He instilled within them a desire for water. For a Darwinian, the match is the product of natural selection. Those creatures in the distant past who were motivated to seek water out-reproduced those who weren't.

More generally, an evolutionary perspective—which I think has considerable advantages over theism in explaining how the mind works—sees the function of pleasure as motivating certain behavior that is good for the genes. As the comparative psychologist George Romanes observed in 1884: "Pleasure and pain must have been evolved as the subjective accompaniment of processes which are respectively beneficial or injurious to the organism, and so evolved for the purpose or to the end that the organism should seek the one and shun the other."

Most nonhuman pleasures make perfect sense from this perspective. When you are training your pet, you don't reward it by reading poetry or taking it to the opera; you give it Darwinian prizes like tasty snacks. Nonhuman animals enjoy food, water, and sex; they want to rest when tired; they are soothed by affection, and so on. They like what evolutionary biology says that they should like.

What about us? Humans are animals and so we share many of the pleasures of other species. The psychologist Steven Pinker notes that people are happiest when "healthy, well-fed, comfortable, safe, prosperous, knowledgeable, respected, non-celibate, and loved." There is quite a bit of pleasure packed into that quote, and I don't doubt for a minute that this is explained through the same process that shaped the desires of animals such as chimpanzees and dogs and rats. It is adaptively beneficial to seek health, food, comfort, and so on, and to get pleasure from achieving these goals. As the anthropologist Robert Ardrey put it, "we are born of risen apes, not fallen angels."

But this list is incomplete. It leaves out art, music, stories, sentimental objects, and religion. Perhaps these are not uniquely human. I once heard from a primate researcher that some captive primates keep security blankets, and there are reports that elephants and chimpanzees can create art (though, as I will discuss later on, I am skeptical about this). But in any case, these are not the usual activities of nonhuman animals. They are entirely typical of our species, showing up in every normal individual. This needs to be explained.

ONE SOLUTION is that our uniquely human pleasures do not emerge through natural selection or any other process of biological evolution. They are the product of culture, and they are uniquely human because only humans have culture (or at least enough culture to matter).

Despite the bad rap that they sometimes get from more adaptation-oriented researchers, those who endorse this sort of culture proposal are not necessarily ignorant or dismissive of evolutionary biology; they don't doubt that humans, including human brains, have evolved. But they disagree with the notion that we have evolved innate ideas, or specialized modules and mental organs. Rather, humans are special in that we possess an enhanced capacity for flexibility, to create and learn biologically arbitrary ideas, practices, and tastes. Other animals have instincts, but humans are smart.

This theory has to be right to some extent. Nobody could deny the intellectual flexibility of our species, and nobody could deny that culture can shape and structure human pleasure. If you win a million dollars in a lottery, you might whoop with joy, but the very notion of money emerged through human history, not due to the replication and selection of genes. Indeed, even those pleasures that we share with other animals, such as food and sex, manifest themselves in different ways across societies. Nations have their own cuisines, their own sexual rituals, even their own forms of pornography, and this is surely not because the citizens of these nations are genetically different.

All of this might tempt someone from a more cultural bent to say that while natural selection plays some limited role in shaping what we like—we have evolved hunger and thirst, a sex drive, curiosity, some social instincts—it has little to say about the specifics. In the words of the critic Louis Menand, "every aspect of life has a biological foundation in exactly the same sense, which is that unless it was biologically possible, it wouldn't exist. After that, it's up for grabs."

I will try to show in the chapters that follow that this is not how pleasure works. Most pleasures have early developmental origins; they are not acquired through immersion into a society. And they are shared by all humans; the variety that one sees can be understood as variations on a universal theme. Painting is a cultural invention, but the love of art is not. Societies have different stories, but stories share certain plots. Taste in food and sex differ—but not by all that much.

It is true that we can imagine cultures in which pleasure is very different, where people rub food in feces to improve taste and have no interest in salt, sugar, or chili peppers; or where they spend fortunes on forgeries and throw originals into the trash; or line up to listen to static, cringing at the sound of a melody. But this is science fiction, not reality.

One way to sum this up is that humans start off with a fixed list of pleasures and we can't add to that list. This might sound like an insanely strong claim, because of course one can introduce new pleasures into the world, as with the inventions of television, chocolate, video games, cocaine, dildos, saunas, crossword puzzles, reality television, novels, and so on. But I would suggest that these are enjoyable because they are not that new; they connect—in a reasonably direct way—to pleasures that humans already possess. Belgian chocolate and barbecued ribs are modern inventions, but they appeal to our prior love of sugar and fat. There are novel forms of music created all the time, but a creature that is biologically unprepared for rhythm will never grow to like any of them; they will always be noise.

ESSENTIAL

Many significant human pleasures are universal. But they are not biological adaptations. They are by-products of mental systems that have evolved for other purposes.

This is plainly true for some pleasures. Many people now get a kick out of coffee, for instance, but this isn't because coffee lovers of the past had more offspring than coffee haters. It is because coffee is a stimulant, and we often enjoy being stimulated. This is an obvious case, but I think that this by-product approach can help explain some of the more difficult puzzles we are interested in. The proposal that I will explore is that these pleasures arise, at least in part, as accidental by-products of what we can call an "essentialist" cast of mind.

One illustration of essentialism comes from a novella by J. D. Salinger, which begins with one of his favorite characters, Seymour, telling a Taoist story to a baby. In the story, Duke Mu asks a friend, Po Lo, to find him someone who can identify a superlative horse. Po Lo recommends an expert, Duke Mu hires him, and soon the expert, Kao, comes back with news of a horse that fits the Duke's requirements, and he describes it as a dun-colored mare. Duke Mu buys the recommended horse, but to his shock, he finds that it is a coal-black stallion.

Enraged, Duke Mu tells Po Lo that this so-called expert is a fool, unable even to appreciate a horse's color or sex. Po Lo, however, is thrilled by this news:

> *"Has he really got so far as that?" he cried. "Ah, then he is worth ten thousand of me put together. There is no comparison between us. What Kao keeps in view is the spiritual mechanism. In making sure of the essential, he forgets the homely details; intent on the inward qualities, he loses sight of the external. He sees what he wants to see, and not what he does not want to see. He looks at the things he ought to look at, and neglects those that need not be looked at."*

The horse, naturally, turns out to be a magnificent animal.

This is a story of essentialism, the notion that things have an underlying reality or true nature that one cannot observe directly and it is this hidden nature that really matters. The classic definition comes from John Locke: the "very being of anything, whereby it is what it is. And thus the real internal, but generally . . . unknown constitution of things, whereon their discoverable qualities depend, may be called their essence."

This is a natural way of making sense of certain aspects of the world. Consider gold. We think about gold, spend money on it, and talk about it, and when we do all of this, we're not thinking and talking about a category of objects that just happen to look alike. If you put gold-colored paint on a brick, it isn't a gold brick. Alchemy is serious business, after all. If you want to know whether something is gold, you need to ask an expert, perhaps a chemist, to do the right test to determine its atomic structure.

Or consider tigers. Most people don't know precisely what

makes tigers tigers, but nobody thinks that it is just a matter of what an animal looks like. If shown a series of pictures in which a tiger is gradually made to look like a lion, even a child knows that it remains a tiger. Rather, the idea is that being a tiger has something to do with genes, internal organs, and so on, invisible aspects of animals that are unchanged by transformations of appearance.

In these examples, people seek out science for the answers, and this makes sense. Scientists are in the business of determining the hidden essences of things. They tell us that there is more than meets the eye, that glass is a liquid, that hummingbirds and falcons are classified together but neither is classified with a bat, and that the genetic connection between humans and chimpanzees is closer than between dolphins and salmon. You don't need to know about science to be an essentialist, though. Everywhere people understand that something might look like an X but really be a Y; they know that a person might wear a disguise or that a food can be prepared to look like something it's not. Everywhere, people can ask, "What is it *really*?"

Social groups are often seen as having essences. So are artifacts, which are objects such as tools and weapons created by people—though here the essences are not physical; they have to do with history and intention. If you want to know what a strange artifact from another time or another country really is, you won't ask a chemist, you would appeal to an expert in archaeology, anthropology, or history.

Essentialism pervades our language. To see this, consider what a nonessentialist language would look like. Jorge Luis Borges invented the Chinese encyclopedia *The Celestial Emporium of Benevolent Knowledge*, which divides animals into categories, including:

Those that belong to the emperor.
Those that resemble flies from a distance.
Those that have just broken a flower vase.

This is clever . . . because it is so weird. The category "Those that resemble flies from a distance" is a logically possible way to group objects, but it's not how we naturally make sense of the world. No real language would have a noun for such a category, because it's too superficial. Real nouns capture something deep; they refer to kinds of things that are thought to share deep properties. As the evolutionary theorist Stephen Jay Gould put it, our classifications don't just exist to avoid chaos, they are "theories about the basis of natural order."

This fact about language makes a real-world difference, particularly when we talk about people. I used to work with children with autism and was constantly reminded to call them "children with autism" instead of "autistics"—the argument being that there is more to these people than their disorder. The noun essentializes; the awkward "children with ______" phrase does not.

It's easy to satirize the political correctness here, but nouns really do carry essential weight. In the movie *Memento*, Leonard Shelby says, "I am not a killer. I'm just someone who wanted to make things right." As he says this, Shelby knows that he killed many people. But that doesn't make him a killer because a killer isn't just someone who has killed; to be a killer is to be a certain type of person, to have a certain deep property, and Shelby denies that this is true of himself. When the baseball player John Rocker was criticized for making a racist remark in an interview, he later said that he isn't a racist: "You hit one home run in the big leagues, it doesn't make you a home run hitter . . . To make one comment like this doesn't make you a racist."

As a milder example, I had dinner with a friend a while ago and she said in passing that she never eats meat. But she bristled when I later referred to her as a vegetarian. "I'm not a fanatic about it," she said. "I just don't eat meat." She saw her diet as an incidental property, not an essential one.

THE TROUBLE WITH ESSENTIALISM

Often essentialism is rational and adaptive: if you attend just to the superficial, you bring home the wrong horse. Someone who looked at the world of plants and animals and didn't realize that members of the same category share deep commonalities—such as docility for certain kinds of animals or curative powers for specific plants—might not live as long as someone with a more essentialist eye. In modern times, the predictive and explanatory victories of science prove that the assumption of a deeper reality is the right one.

But essentialism sometimes leads us to confusion. The social psychologist Henri Tajfel began a classic line of research into "minimal groups." He found that if you separate people into groups based on the most arbitrary of considerations—in some studies, literally a coin flip—people will not only favor their own group but will also believe that there are significant differences between the groups, and that their group is, in an objective sense, superior. The essentialist bias leads us to see deep commonalities even when none exist.

It is not surprising, then, that when the differences are blatant, such as in facial shape or skin color, we don't brush them off as arbitrary variation; we think they matter. And to a point, they do. If you know what someone looks like—for instance, what color his skin is—you now are prepared to guess many invisible facts about him, such as his relative income, religion, political affiliation, and so on. (As I write this, an American with dark skin color is far more likely to vote for a Democrat than an American with light skin color.) Race matters in large part because people who look different come from different countries, settle in different neighborhoods, and have different histories.

But our essentialism goes beyond this; people tend to think about human groups, including races, in biological terms. The psychologist Susan Gelman tells of someone who claimed, "I can't date anyone who's not a mitochondrial Jew." Mitochondrial DNA is passed down the maternal line, and this was a clever way of stating a certain definition of Judaism, but it nicely captures how we think about human groups in a biological manner. Before DNA, it used to be blood, as in the notion that a single drop is sufficient to count someone as being of African descent.

Biological essentialism toward race is not *entirely* mistaken. Swedes are larger than Japanese who are larger than pygmies, and some of this is plainly due to genes. And when we classify ourselves as falling into one category or another, even the most liberal and determinedly antiracist people understand that this is a question about biological origin. The psychologist Francisco Gil-White points out that when someone says that she is half Irish, one-quarter Italian, and one-quarter Mexican, she's not talking about the extent to which she has mastered different cultures or which groups she has decided to affiliate with—she is talking about the ethnicities of her grandparents.

But the categories aren't as real as some people think they are.

Genes do not determine whether someone is Jewish, for instance. An adult can become a Jew by conversion; a child can become a Jew by being adopted into a Jewish family. My children are the offspring of a Jewish father and a non-Jewish mother—are they Jewish, half Jewish, or not-at-all Jewish? The answer is a political or theological one, not a scientific one. Maybe this is an obvious case, but the same point holds more generally. Consider that President Barack Obama is usually described as African American or black, even though he is the offspring of one parent who would typically be defined as black and the other who would typically be defined as white. Given the social context, the black trumps the white. More generally, categories such as "black" include people from radically different groups—from Haitian to native Australian—who are thrown together by dint of a shared property that is literally only skin deep. To think that they share a deeper connection is essentialism run amuck.

THE ESSENTIAL CHILD

Susan Gelman begins her wonderful book *The Essential Child* with the story of when she was four or five and asked her mother how boys and girls were different. Her mother said, "Boys have penises, girls don't." Gelman was incredulous. "Is that all?" she asked. Given how differently boys and girls dressed and acted and played, she was looking for something more interesting, something deeper. The point of her story is to out herself as a child essentialist, a preamble to her argument that all children are essentialist.

This is admittedly a controversial claim within the field of psychology. The dominant view, established by the Swiss developmental psychologist Jean Piaget, and defended by some noted scholars today, is that children start off with a superficial orientation toward the world, limited to what they can see, hear, and touch. From this perspective, essentialism has historical and cultural origins. In the physical and biological domains, it was a discovery, an intellectual accomplishment achieved first by philosophers and later by scientists. Most people would never have figured it out for themselves. The philosopher Jerry Fodor states: "*of course* Homer had no notion that water has a hidden essence, or characteristic microstructure (or that anything else does)." We learn about this in school. In the domains of race and sex and caste, essentialism is a myth invented by the powerful to convince people that these social categories are natural and immutable.

We are far from a complete theory of the origin of essentialism. But I think the evidence is now abundant that much of essentialism does not have cultural origins. It is a human universal. Homer probably *did* think that water has an essence.

Much of the research here comes from developmental psychology. We know that even babies can infer invisible properties based on what things look like. If nine-month-olds find that a box makes a sound when you touch it, they expect other boxes that look the same to make the same sound. Older children do more; they make generalizations based on the category something belongs to. In one study, three-year-olds are shown a picture of a robin and told that it has a hidden property, such as a certain chemical in its blood. Then they are shown two other pictures: one of an animal that looks similar but belongs to a different category, such as a bat; the other of an animal that looks different but belongs to the same category, like a flamingo. Which one has the same hidden property? Children tend to generalize on the basis of category, choosing the flamingo. This doesn't show that they are fully essentialist, but it does show that they are sensitive to something deeper than appearance. Other studies using modified procedures have found the same effect with children before their second birthday.

Other experiments find that young children believe that if you remove the insides of a dog (its blood and bones), it isn't a dog anymore, but if you remove its outside features, it still is. And children are more likely to give a common name to things that share deep properties ("have the same sort of stuff inside") versus those that share superficial properties ("live in the same kind of zoo and the same kind of cage").

My colleague at Yale, Frank Keil, found some of the most striking demonstrations of child essentialism. He showed children pictures of a series of transformations: a porcupine surgically transformed so as to look like a cactus, a tiger stuffed into a lion suit, a real dog made to look like a toy. The neat finding is that children rejected such radical transformations as changing the category—regardless of what it looks like, it is still a porcupine, a tiger, or a dog. Only when the children were told that the transformations occured on the inside—the innards of these creatures were changed—could they be persuaded that these transformations lead to a real change in category.

Like adults, young children expect names to refer to objects that share deep hidden properties. Susan Gelman once showed her 13-month-old son a button on her shirt and called it "button." He then started to press it, because, though it didn't look much like a button on his electronic toys, he knew what category it belonged to, and that's what you do with a button. For older children, you get the same subtle appreciation of the force of a noun that you find in adults. One four-year-old made the point when describing a violent playmate: "Gabriel didn't just hurt me! He hurt other kids, too! He's a *hurter*! Right, Mom? He's a *hurter*!" The child is presumably stressing that this sort of behavior reflects a deeper aspect of Gabriel's nature. And in their experimental work, Gelman and Gail Heyman told five-year-olds about a child named Rose who often eats carrots, and added, for half of the children: "She is a carrot-eater." This name has an effect; it caused those children to think of Rose as a more permanent eater of carrots—she will eat them in the future, even if her family discourages it. It is part of her nature.

Some scholars have argued that child essentialism arises from a specialized system that's just for thinking about plants and animals. But in my own work, I find that children are highly essentialist about everyday artifacts. When they hear a name used to refer to a novel human-made creation, they extend that name to objects that were created with the same intention, regardless of what they look like.

Also, children are essentialist about categories of people. In fact, one of the strongest examples of essentialism concerns the difference between the sexes. Before ever learning about physiology, genetics, evolutionary theory, or any other science, children think that there is something internal and invisible that distinguishes boys from girls. This essentialism can be explicit, as when one girl explained why a boy will go fishing rather than put on makeup: "'Cause that's the boy instinct." And seven-year-olds tend to endorse statements such as "Boys have different things in their innards than girls" and "Because God made them that way" (a biological essence and a spiritual essence). Only later in development do children accept cultural explanations, such as "Because it is the way we have been brought up." You need to be socialized to think about socialization.

This research is ongoing, but there is an emerging consensus that children are natural-born essentialists. The scope of this essentialism is broad; we attribute essences to animals, artifacts, and types of people.

LIFE FORCE

I have described essentialism so far as a way of thinking about categories. It is the notion that there is something deep within each tiger, say, that makes it a tiger. But now consider the view that there is an essence within each individual that makes it special: not tigers versus lions, but this tiger versus that tiger.

The capacity to think of specific individuals is a significant aspect of mental life, and it extends to the most uninteresting things. The philosopher Daniel Dennett gives the example of someone carrying a penny with him from New York to Spain and impulsively tossing it into a fountain. It now lies with the other pennies, and there is no way that he could ever tell it from the other pennies, but, still, he appreciates that one and only one of the pennies is his. If he were to scoop up a penny from the fountain, it would either be the one he brought from New York or a different penny.

Thinking about individuals is a significant cognitive ability, but it is not essentialism. You can understand that the pennies each have their own histories, but this doesn't mean that they contain anything more, anything that one could think of as an essence.

But we do think that some individuals have their own essences. This is particularly the case for people or objects that are closely related to people. In many cultures, these essences are understood in terms of some invisible force. The psychologists Kayoko Inagaki and Giyoo Hatano argue that children start off as "vitalist"—

they assume that living beings have an animating force inside them. Such a belief is common across societies, as "chi," "ki," "elan vital," "mana," "life force"—or "essence." It is thought of as part of a person, some have it more than others, and it can be passed from people to objects and then back again. The anthropologist Emma Cohen told me about her research into Axe (pronounced ah-shay) in the Afro-Brazilian religion:

> *The people I chatted with explained how mere things, artifacts and everyday objects can become sacred via Axe-giving rituals. It is also present in all humans to various degrees, and can be "topped up" through participation in rituals. Having it expresses power. When you are sick, for instance, you should seek healing from someone with greater Axe. And since you can't tell simply by looking who has more and who has less, you might blame the failure of a ritual due to weak Axe, which people do. Some religious houses have more Axe than others and Afro-Brazilian religionists say that when you're in a house with more Axe, you feel better.*

This is an example of how the life force is involved in religious ritual, but it shows up in our secular lives as well. We seek out contact with special people. A mere thing that has been touched by a special person gains value, which is one reason why people pay a lot for objects such as JFK's tape measure. Indeed, as described in a later chapter, my colleagues and I find that people will pay dearly for an admired person's sweater (such as George Clooney's)—but the price drops if it has been sterilized, because this obliterates the essence.

Then there is contact with the actual person. Sometimes it can be affecting just to be looked at by a high-status individual. In an intriguing discussion, the writer Gretchen Rubin connects this experience to the notion in Hindu philosophy of *darshan*, a Sanskrit term meaning "sight." This can be draining for the person who is thought to be giving off the energy, so much so that some celebrities have contracts prohibiting their employees from making eye contact with them.

Better than a look is a pat on the shoulder, and better than that is a handshake. Expressions such as "I won't wash my hand for a week" capture the notion that there is some remnant of the famous person on your hand, one that you don't want to lose. More intimate than a handshake is sexual intercourse, which is one of many reasons why the powerful have little problem finding sexual partners.

You can get more physically intimate than sex, though. Consider the overheard phone conversation in which Prince Charles expresses the desire to be reincarnated as his mistress's tampon—a desire that is both creepy and sweetly romantic. There is carving up the body of that special someone and eating it, in the hopes that you yourself will now get the person's powers, a practice we will turn to in the next chapter. And there is organ transplantation, in which one person comes to possess a part of another, a particularly intimate act—the ethicist Leon Kass once described it as "a noble form of cannibalism." Indeed, many believe that the recipients of transplants take on the properties of the donors.

There are differences between the category essentialism we started off with and this sort of life-force essentialism: category essences are thought to be permanent and immutable, while life-force essences can be added and subtracted and passed on. What they have in common is that they are invisible, they can determine what an object is, and they can matter a great deal.

One example of how essentialism matters is based on eyewitness accounts of the search for the 14th Dalai Lama. The relevant section concerns the testing of a particular two-year-old boy in his remote home village. A group of bureaucrats brought with them the belongings of the late 13th Dalai Lama, along with a set of inauthentic items that were similar or identical to these belongings. When presented with an authentic black rosary and a copy, the boy grabbed the real one and put it around his neck. When presented with two yellow rosaries, he again grasped the authentic one. When offered two canes, he at first picked up the wrong one, then after closer inspection he put it back and selected the one that had belonged to the Dalai Lama. He then correctly identified the authentic one of three quilts. As a final test, the boy was presented with two hand drums: a rather plain drum (authentic) and a beautiful damaru far more attractive than the original. That is, there was a forced choice between an uninteresting object with the essential property versus a highly salient distracter. Here is the report of what they found: "Without any hesitation, he picked up the drum. Holding it in his right hand, he played it with a big smile on his face; moving around so that his eyes could look at each of us from close up. Thus, the boy demonstrated his occult powers, which were capable of revealing the most secret phenomena."

Another observer described this recognition ability as a sign of "super-human intelligence." (Note that the use of exact copies means that the boy could not succeed through past-life memory; some special power to discern invisible essences would be required.) The point here is not that the authentic objects were actually imbued with the essence of the 13th Dalai Lama; what matters is that the Tibetan bureaucrats believed that they were and constructed a procedure that presupposes the existence of invisible essences—essences that require special powers to perceive—and used this procedure to make an important decision. The boy become the 14th Dalai Lama, Tenzin Gyatso.

SMARTER THAN WE LOOK

In the chapters that follow, I will argue that the pleasure we get from many things and activities is based in part on what we see as their essences. Our essentialism is not just a cold-blooded way of making sense of reality; it underlies our passions, our appetites, and our desires.

There is a lot going on in psychological essentialism, different notions of essences being explored. There is category essentialism and life-force essentialism; there is the physical essences of natural things like animals and plants, and the psychological essences of human-made things such as tools and artwork. My own attempt to extend essentialism to pleasure is going to be correspondingly broad. At times I will be relating pleasure to category essences of the standard sort, such as in the discussion of sex, where categories such as *male*, *female*, and *virgin* turn out to be highly relevant. Sometimes the essence is more similar to the invisible life force, as when we discuss how certain consumer products get their value. Sometimes the focus will be on the role of inferred internal structure, as with our taste for bottled water; sometimes it will be on human history, as with our experience of paintings and stories. And the book will end with a discussion of the more general intuition that there is an underling reality that transcends everyday experience, an intuition that might be at the foundation of the pleasure we get from both religious practice and scientific inquiry.

This is admittedly a complicated take on pleasure. So be it: people are complicated critters. We often miss this complexity. Certain facts about our psychology are so immediate and obvious that it's hard to see them as requiring any explanation at all. William James made this point with typical eloquence in 1890:

> *To the metaphysician alone can such questions occur as: Why do we smile, when pleased, and not scowl? Why are we unable to talk to a crowd as we talk to a single friend? Why does a particular maiden turn our wits upside-down? The common man can only say:* "Of course *we smile,* of course *our heart palpitates at the sight of the crowd,* of course *we love the maiden."*

He goes on to explain how these feelings are accidental properties of an animal's makeup:

> *And so, probably, does each animal feel about the particular things it tends to do in the presence of certain objects . . . To the lion it is the lioness which is made to be loved; to the bear, the she-bear. To the broody hen, the notion would probably seem monstrous that there should be a creature in the world to whom a nestful of eggs was not the utterly fascinating and precious and never-to-be-too-much-sat-upon object which it is to her.*

When it comes to pleasure, it is tempting to attribute our reaction to a thing to properties of the thing itself. *Of course* we are tongue-tied by the maiden—she looks so darn hot. How could she not turn our wits upside down? *Of course* we are enraptured by a tiny baby—it's so adorable.

The depth of pleasure is hidden from us. People insist that the pleasure that they get from wine is due to its taste and smell, or that music is pleasurable because of its sound, or that a movie is worth watching because of what's on the screen. And of course this is all true . . . but only partially true. In each of these cases, the pleasure is affected by deeper factors, including what the person thinks about the true essence of what he or she is getting pleasure from.

2

FOODIES

IN 2003, ARMIN MEIWES, A 42-YEAR-OLD COMPUTER EXPERT, went online looking for someone to kill and eat. After several interviews, he chose Bernd Brandes. The two men met one night in Meiwes's farmhouse in a small town in Germany. They talked for a while, and Brandes took several sleeping pills and finished off half a bottle of schnapps. Meiwes then cut off Brandes's penis and fried it in olive oil. The two men tried to eat it, without success. Meiwes read a Star Trek novel, and Brandes, bleeding heavily, lay in a bath. A few hours later Meiwes killed Brandes by stabbing him in the neck with a kitchen knife, kissing him first.

Meiwes then chopped Brandes up and stored him in the freezer, next to some pizza. In the weeks that followed, he defrosted and cooked pieces of Brandes in olive oil and garlic, devouring about 44 pounds of him. He used his best cutlery, lighting some candles and accompanying his meals with a South African red.

This episode is interesting in several regards. For one thing, although the act was consensual, many people believe that Meiwes did something terribly wrong. He was first convicted of manslaughter and later, when the prosecution appealed, found guilty of murder. Articulating just what it is about this act of consensual cannibalism that people find so immoral—including people of liberal inclination, who are prone to believe in human autonomy and freedom and would usually agree that people should be free to do whatever they want so long as it doesn't infringe on the will of others—might give us some insight into moral reasoning and moral principles.

There is also the clinical question of why Meiwes developed a taste for human flesh. As one would expect for a creature of our time, he had his own psychological story—his father abandoned him, he was lonely, he fantasized about having a younger brother whom he could keep forever by eating him. This idea of fidelity through consumption seems to be a common theme in such cases. One expert had a similar explanation for Jeffrey Dahmer, the American cannibal killer, arguing that Dahmer ate his lovers because he wanted them never to leave him.

(And what about Brandes? I can understand wanting to die, but who would want to be killed by a stranger who plans to eat your corpse? Brandes was not unique in his interest—about two hundred men responded to Meiwes's ad on the Internet. This was how he got caught; a student, surfing the Web, monitored these discussions and notified the authorities.)

Still, what does this story have to do with the everyday pleasure of food? Societies in which people eat people are rare, so rare that some have doubted that they have ever existed. Cannibal killers are more common in horror movies than in reality. When Dahmer was interviewed in prison, he plaintively asked the doctors whether there was anyone else in the world like him.

There are two reasons why cannibalism is a good place to begin a discussion of the pleasure of food. First, it provides a useful way to approach the question of why some things are good to eat and others are not. Exploring why we feel so strongly that people don't fall into the food category might give us insight into more usual likes and dislikes. Second, the psychology of the cannibal turns out to reflect an extreme version of what normal people think about the foods that they normally eat. It illustrates essentialist belief in its sharpest form.

By eating Brandes, Meiwes believed that he was doing something more than merely consuming protein and fat; he was consuming Brandes's *essence*. He insisted that there were psychological benefits to devouring a person. He felt more stable afterward and incorporated some of Brandes's qualities: "With every bite, my memory of him grew stronger." Brandes was fluent in English and Meiwes claimed that since eating him, his own English improved. This notion of incorporating a person's essence was captured by a song inspired by this event performed by a German metal band. The chorus begins, "Denn du bist was du isst," which means "Because you are what you eat."

PICKY

When I first became interested in the pleasure of eating, I assumed that the explanation for why we like some foods and not others was going to come from physiology and evolutionary biology. We would explain what humans eat from the study of taste and smell, from the anatomy of the senses. We should be able to predict the foods that we like from facts about what our bodies most need and the environments in which our species has evolved. Our taste in art and music might plausibly be the result of culture, temperament, experience, and luck, but surely taste in food is a biological matter, shaped by our species' history.

This isn't entirely wrong. There do exist some hard-wired preferences. Humans naturally like sweet things, because sugar is a good source of calories, and we dislike bitter things, because bitterness is a cue to toxicity. Some foods, such as chili peppers, cause an unpleasant "burn"; mothers in some cultures put chili on their breast so as to begin the weaning process, and it would be cruel to squirt Tabasco sauce into a baby's mouth.

But that is about it for human universals. As the psychologist Paul Rozin has pointed out, we are omnivores—we eat just about anything that we can digest. Relative to other animals, there are few biological constraints on the human diet.

What about human differences? Some of this can be explained genetically. Most people in the world are lactose intolerant; milk is bearable for only a minority of humans. One fascinating discovery is that there is more than one type of tongue—about one-quarter of us have a high density of receptors and are, as Linda Bartoshuk originally put it, *supertasters*. You can find out whether you are one by putting blue food coloring on your tongue, and then asking a friend to count your still-pink taste buds. Your fungiform papillae, where the taste buds lie, don't absorb the coloring. As a simpler procedure, you can get hold of some paper with 6-n-propylthiouracil (PROP) on it (accessible online) and stick it in your mouth. If it tastes like paper, you are like most people; if it is unpleasant and bitter, congratulations, you are a supertaster!

Supertasters are more likely to dislike whiskey and black coffee, Brussels sprouts and cabbage. They are especially sensitive to the acidity of a grapefruit and the burn of chili powder. But while supertaster status is related to food preferences, it's an imperfect predictor. My wife, a supertaster, has the predictable behavior of disliking beer and diet soft drinks, but she enjoys bitter vegetables like broccoli rabe. It's surprisingly difficult to read off taste preferences from facts about our physiology.

A few years ago, there was discussion of tongue physiology among wine experts, sparked by a symposium in Napa, California, at which attendees were given the PROP test. Predictably, those wine tasters who passed boasted about their supertaster status. What complicates the issue, however, is that there is no evidence that supertasters, despite the *super-* in the name, are better at discriminating different flavors than the rest of us. In fact, they tend to be *less* prone to enjoy wine, given their dislike of astringency and acidity.

Nobody can yet explain most of the variation in food preferences. You can take siblings who are raised together in the same house and who share half of their genes and still there are differences. I hate cheese, my sister loves it, and I have no explanation why.

Still, there are some factors that do make a difference. If you want to know what someone likes to eat, the best question is: Where do you come from? Culture explains why some people enjoy kimchi, others tortillas, others Pop-Tarts. It explains why Americans and Europeans don't eat bugs, rats, horses, dogs, or cats, while others enjoy them. Some even eat human flesh, though under certain restricted circumstances. All of this is best explained by where they come from and how they were raised.

We can now pass the buck to a sociologist or anthropologist, asking about the forces that cause societies to establish certain tastes. The anthropologist Marvin Harris has developed a well-known approach along these lines, based on optimal foraging theory. For Harris, there is a logic to these choices. Some foods just aren't worth the trouble of eating. Americans don't eat dogs, for instance, because they are worth more alive—they offer companionship and protection. Bugs aren't lovable, but they are time-consuming to collect; not worth the effort. (The exceptions are those that are large, or swarm together in high-density clumps, or are worth killing because they are bad for crops; accordingly, bugs like locusts are sometimes fine to eat—John the Baptist is described as surviving in the wilderness on nothing but locusts and honey.) In places that do not eat cows, it turns out that cows are worth more alive than dead.

While the specifics of these proposals are controversial, Harris is likely right that such restrictions are not accidents. But the problem from the psychologist's standpoint is that there is no obvious connection between the cultural explanation and the psychological one. Harris's theory doesn't explain the food preferences of individuals. I was raised in Canada, and no doubt Harris could provide an elegant account of why Canadians don't eat rats, but this doesn't explain why I personally avoid rats. Rational considerations might determine cultural choices; they don't shape individual tastes. I might be convinced that rat meat is nutritious, healthy, and (to an unbiased taster) yummy, but, still, having a plate of fried rat placed in front of me would make me gag. Conversely, I have been entirely persuaded that there are excellent moral and practical reasons not to eat cows. But steak still tastes delicious.

This is typical of cultural learning—the explanation at the cultural level usually has nothing to do with the explanation at the personal level. There are historical reasons why people in Damascus tend to speak Arabic and people in New Haven tend to speak English, or why Damascus residents are likely to be Sunni Muslim and New Haven residents are likely to be Christian. These are not random events; they have historical explanations. But children raised in these cultures don't know these historical facts when coming to speak their language and worship their god.

So what determines individual preferences? A promising direction is to look at personal experience. Humans and other animals have special neural systems that ward us away from foods that are bad for us. If you eat a novel food and later become sick or nauseous, you will avoid that food later—the very thought of eating it will turn your stomach. When I talk about food in my Introduction to Psychology class, I ask for stories about food aversions, and there are always some people who cannot eat something because they got sick while first trying it. For one student, it was eating sushi as she was coming down with the flu. For me, it was mixing ouzo—a Greek liquor—with beer as a high school student and becoming violently ill. For years later, I would be sickened by the distinctive licorice smell.

Another sort of learning is through observation of others. Perhaps, like rat pups, we figure out what foods are safe to eat—and hence which foods we should get pleasure from—by monitoring what our parents give us to eat and observing what they eat themselves. Parents share the children's environments, and tend to love their children and care about their welfare, so it seems like a perfectly reliable learning mechanism.

Oddly, though, for humans it's not that simple. It turns out that there is only a small relationship between the preferences of parents and those of their young children. There is evidence for a stronger relationship between siblings, as well as between married couples. This last finding is particularly puzzling, since you are usually not genetically related to your spouse.

One can explain these facts by taking seriously the idea that food learning is in part a form of cultural learning. It is more than ascertaining what is nutritious and nonlethal. It is part of being socialized into a human group. And social learning, as the psychologist Judith Harris and others have emphasized, is accomplished by attending to one's peers. You don't eat like your parents for the same reason you don't dress like your parents, or swear like them, or enjoy the same music. This explains that lack of relationship between parent and child, and it explains as well the close tie between sibling and sibling, and between husband and wife.

For the youngest of babies, there is no choice but to attend to adults. Still, babies are smart enough to engage in some social reasoning. In one clever study, American 12-month-olds watched as two unfamiliar adults each ate a strange food. The two strangers spoke to the babies, one of them in English, the other in French. When later asked to choose between the two foods, these American babies preferred the food eaten by the English speaker, reflecting a tendency to learn from a person who is more similar to them.

DISGUSTING

The problem with human flesh is not that it tastes bad in some objective sense. By all accounts, if you like pork, you would be perfectly comfortable eating a person, so long as you didn't know what you were eating. (It has been claimed that the closest thing to eating a person is eating the commercial product Spam.) Indeed, there are many stories, riddles, and fables that assume that one can be tricked into eating human flesh and liking it, only later discovering what it is.

What's wrong with human flesh is how we think about it. Marvin Harris nicely makes this point regarding insects: "The reason we don't eat them is not that they are dirty and loathsome; rather, they are dirty and loathsome because we don't eat them." Similarly, what bothers us about human flesh is that we know what it is. It is loathsome. It is disgusting.

The emotion of disgust plays an interesting role in what we like to eat. Disgust has evolved as an aversion to rot and contamination, and particularly the risk of rotting meat. One might dislike, for instance, yams, apple pie, licorice, baklava, raisins, or whole-wheat pasta, but there is typically a stronger reaction to meat, to eating dog, horse, and rat. The strong nonmeat aversions that do exist tend to prove the rule—they tend to be derived from animals (like cheese or milk) or be foods that resemble animals in appearance or texture (Rozin notes that shellfish are often seen as resembling genitalia).

Charles Darwin expressed our reaction to new meat in unusually strong terms: "It is remarkable how readily and instantly retching or actual vomiting is induced in some persons by the mere idea of having partaken of any unusual food, as of an animal which is not commonly eaten; although there is nothing in such food to cause the stomach to reject it." This is admittedly extreme; either Darwin is exaggerating or his Victorian contemporaries were particularly fragile—I don't know anyone who would vomit at the mere thought of eating an unusual animal. He is surely right, though, that it is gross.

The developmental story of disgust is one I told in detail in my last book, *Descartes' Baby*. Here is the abbreviated version: Babies and young children cannot be disgusted. They don't mind their own waste products, or anyone else's for that matter. They'll eat grasshoppers and other bugs. Paul Rozin and his colleagues did an experiment in which they offered young children dog feces to eat (it was actually a combination of peanut butter and smelly cheese). They gobbled it up. As best I know, no psychologist has ever given a toddler a hamburger and described it as human flesh, but I bet that the child would cheerfully wolf it down.

Disgust kicks in at roughly the age of three or four. Children will then veer away from feces and urine, and they know that a glass of juice or milk with a cockroach in it isn't fit for drinking. Sometimes they are hypersensitive, obsessively concerned about what their food touches and where it has been. William Ian Miller, in *The Anatomy of Disgust*, talks about his fastidious children: his daughter who refused to wipe herself at the toilet because she was afraid to soil her hand and his son who would remove his pants and underpants if a drop of urine went astray.

Nobody knows what triggers the emergence of disgust. The Freudian notion that it is linked to toilet training is not plausible. There are huge social differences in how children are taught to urinate and defecate, and many cultures have no toilets at all. Nevertheless everyone, everywhere, is grossed out by urine and feces. A further problem for Freud is that blood, vomit, and rotten flesh are all universally disgusting, but we certainly don't learn about these through toilet training. It seems more likely that what drives the emergence of disgust is biological timing, part of neural development.

Some substances such as feces are universally repugnant, but there is also cultural variation, particularly regarding our response to meat. Darwin's observation tells us something important about how this learning takes place. It is not that children learn, one-by-one, which meats are disgusting. Rather, meat is guilty until proven innocent. That is, children monitor the sorts of flesh that the people around them eat, and they grow to be disgusted at everything that isn't consumed. Meat is special in this regard. An adult might be willing to eat new fruits and vegetables or other foods—when I was a child, I never ate granola bars, California rolls, shrimp dumplings, or crab cakes, but I like them all now. I wouldn't even try rat or dog.

Some of the research on this topic has been done by the military, because soldiers, and particularly pilots, might find themselves in situations in which their preferred diet is not available. Getting people to eat disgusting things is also a perfect way to study their receptivity to taking orders.

This was the motivation for a study published in 1961 by Ewart E. Smith, which begins with this faintly ominous sentence: "The Army Quartermaster recently presented the Matrix Corporation with the problem of determining the best methods for changing attitudes in military organization." So they explored different techniques for getting people to eat disgusting foods, including bugs, fried grasshoppers, and "irradiated bologna sandwiches." The main finding is that you can make people eat these things, but you can't make them like them.

WHY DO PEOPLE EAT HUMAN FLESH?

People will eat human flesh out of desperation and hunger, but one of the nastiest things to say about people is that they enjoy the cannibal lifestyle by choice. In 1503, Queen Isabella ruled that the Spaniards could only take as slaves those whose lot would be improved by enslavement, which motivated Spanish explorers to tell lurid stories about other cultures. And what could be worse than cannibalism? Noting the powerful stigma, one scholar in the 1970s wrote a book arguing that there are no such things as cannibal cultures; it's all a myth.

Some accusations are true, however, and the evidence is now overwhelming that such societies exist. It would be strange if it were otherwise. From an evolutionary perspective, life is a competition for, among other things, protein. Being in a rich industrial society, it is easy to forget that most humans have lived most of their lives desperate for more meat. It must have been obvious that the solution to this problem was right in front of them, in their children, friends, neighbors, and certainly in those that they hated. Certainly other primates have figured this out; a major cause of death for baby chimpanzees and gorillas is infanticide. This is for many reasons, but one consideration, as the anthropologist Sarah Hrdy puts it, is that the babies are a "delectable source of proteins and lipids."

There are two ways to be a cannibal, each with distinct advantages and disadvantages and each presupposing that you are imbibing the essence or spirit of the person being eaten.

Option 1: Endocannibalism: Wait until people die of natural causes. Then eat them.

On the positive side, this is light work. It does not require effort or violence. The negative side is that your meals tend to be old, desiccated, and often ridden with dangerous diseases. In 1976, Carleton

Gajdusek won the Nobel Prize in part for his finding that the disease of kuru among the Fore people in Papua New Guinea is the result of their cannibalistic practices, in particular, their eating of brains.

If you're an endocannibal, there are many ways you might eat your dead. Sometimes it's solemn; sometimes giddy. On rare occasions, people eat the whole corpse, but typically the flesh isn't consumed; rather the bones are ground up, or the body is burned to ashes, and then the powder is mixed with a drink or with something like mashed banana. The rock star Keith Richards described a modern variant in an interview in the British music magazine *NME*:

> *The strangest thing I've tried to snort? My father. I snorted my father. He was cremated, and I couldn't resist grinding him up with a little bit of blow.*

The point is not the ingestion of protein. The idea is to take in the essence of someone you love. For endocannibals, failure to do so might mean poor health, infertility, or weak children.

Option 2: Exocannibalism: Find young and healthy people from other groups. Kill and eat them.

This has the advantage that young healthy people are fine sources of protein, and the disadvantage that they don't want to be eaten and are highly motivated to take steps to avoid this fate, steps that can be dangerous to the cannibal wannabe.

Some people eat their prisoners. This is typically a violent event, and this violence reflects certain essentialist beliefs. The prisoners might be forced to fight, in the hopes that their bravery would pass into the bodies of the people who will eat them. The Aztecs, for instance, tied their prisoner by the waist, gave him a weapon, and repeatedly attacked him until he fell. Then he was stretched out and his body was flayed so that the skin could be used as a cloak, and his flesh carved and eaten. Some societies have elaborate rituals, including prepared dialogues between the cannibals and their captors. In one report in Brazil in 1554, the dialogue goes as follows:

Tribesman: I am he that will kill you, since you and yours have slain and eaten many of my friends.

Prisoner: When I am dead I shall still have many to avenge my death.

Both types of cannibalism, then, have, as one motivation, the appropriation of others' spirits, their essences. Is this the real reason why people eat people? A cynic might wonder if the ritual emerges from some other reason and then these essentialist beliefs are tacked onto it, in the same way that some people who follow kosher law talk about the health benefits of such a diet, even though this isn't the original motivation for their choice.

Exocannibalism may indeed have started because of the health benefits of eating healthy humans with the added bonus of terrorizing your enemies. But it is not plausible for endocannibalism. Grinding up old people and eating them has no tangible benefit. It is better to take the cannibals at their word: they eat them to preserve and protect their loved ones' invisible essences.

• • •

EVERYDAY CANNIBALISM

The discussion of cannibalism so far has centered on the exotic, the primitive, and the criminally insane. You are probably not one of those, and hence you are probably not a cannibal. But you are likely to do cannibal-like things and think cannibal-like thoughts. The notion that you can acquire someone's essence by ingesting him or her is commonplace.

One well-known example of this concerns the Eucharist, a ritual that millions of Catholics regularly practice, in which they describe themselves as ingesting the body and blood of Christ. The cannibalistic association here is hard to miss and was used in attacks against the Catholics in the sixteenth century, when people argued that this ritual reflected a more general propensity to eat human flesh. This is itself reminiscent of the blood libel directed toward the Jews, who were said to cook Christian babies and use them for matzo. There is rich theological debate over whether the Eucharist really counts as cannibalism, but regardless, it is certainly cannibal-*like*:

> *He who eats my flesh and drinks my blood has eternal life, and I will raise him up on the last day.*

I'm not Catholic myself, but there's something that makes sense about this, about taking in the essence of someone by eating him or her. It is a loving act, reminiscent of the monsters in Maurice Sendak's *Where the Wild Things Are*. As the boy, Max, starts to return home, they cry: "Oh please don't go—we'll eat you up—we love you so!"

The only example that I know about of contemporary, socially approved, honest-to-God cannibalism (not symbolic, but real flesh and blood) is the eating of placentas. This is more common in parts of Asia, but it does exist in the United States and Europe, where it is partially motivated by the New Age movement. One Web site discusses this in the context of "solidarity with other mammals" and describes various recipes:

> *The most popular method, it seems, is to prepare the placenta fresh with garlic and tomato sauce. It can also be made into a lasagna or a pizza, folded into a vegetable-juice cocktail or a placenta smoothie, or dried and sprinkled in a salad. At the cutting edge of placenta cuisine is placenta sashimi and placenta tartar (a breeze to prepare—just slice and serve!).*

It is said, correctly, that the placenta is a good source of protein, but there is no shortage of protein in the lifestyle of a modern American. This is not why some people go to the trouble of eating it. Rather, the placenta is sometimes said to have certain powers, such as immunization against postpartum depression.

There has been at least one televised case of placenta eating. In a 1998 episode of the British series *TV Dinners*, the celebrity chef devised a surprise dinner for someone who had just had a child. He made the placenta into a pâté and served it on focaccia. Many of the dinner guests were shocked and the show was severely reprimanded by the British Broadcasting Standards Commission.

This is harmless fun, perhaps, but there do exist terrible manifestations of modern cannibalism. There is trafficking in human body parts, particularly of the young, as part of the African belief system known as *muti*. In Tanzania, witch doctors market skin, bones, and hair of albinos as part of potions that are thought to provide good fortune. Dozens of albinos have been killed, including several young children.

YOU ARE WHAT YOU EAT

An essentialist mind-set might make you stop eating certain foods. When Gandhi first ate goat, he claimed to feel that the soul of the animal was crying out in his belly, an excellent impetus to vegetarianism. Essentialism might make you eat more of certain foods. Before the invention of Viagra and its offshoots, desperate men would eat animals and animal parts as aphrodisiacs. The specific meal would be chosen for different reasons, sometimes because of the youth of the animal, sometimes its virility, sometimes because the part represented the aspired-for erect penis, and sometimes for no obvious reason at all. Some presumed impotence cures were:

- human body parts
- rhino horns
- tiger penis
- seal penis
- oysters
- prawns
- crocodile teeth
- roasted wolf penis

Meat is said to be good for this sort of thing, and, in unpublished work, Paul Rozin has argued that humans everywhere associate meat with manliness. When I was a graduate student I had a

Russian roommate who would insist on the relationship between meat-eating and sexual virility, and would deride the potency of his vegetarian friends.

A quite different effect is associated with water. Americans spend about $15 billion a year on bottled water, more than we spend on movie tickets. We drink more bottled water than milk, coffee, or beer. This is puzzling, because in most parts of the country, bottled water is no healthier or tastier than tap water (it is often worse). There is also the considerable environmental cost of the production of plastic bottles and the transportation of the water by truck. And bottled water costs more, by volume, than gasoline. What makes it so appealing?

One answer is that we are drawn to its purity. In general, people prefer natural over artificial. We are wary of medical antidepressants but comfortable with herbal remedies such as ginkgo biloba. Genetically modified foods are repellent to many. This hunger for the natural poses a problem from the standpoint of marketing. As the writer and activist Michael Pollan explains in *The Omnivore's Dilemma*, it is hard to make money from whole natural foods. This is in part because, as a vice president for General Mills pointed out to him, you can't easily distinguish your company's corn or chicken from everyone else's corn or chicken. To turn a profit, it helps to make the corn into a brand-name cereal and the chicken into a TV dinner. Pollan describes how in the 1970s, a food additive manufacturer called International Flavors & Fragrances hoped to dissuade people from natural foods, arguing that the artificial is better for you. Natural ingredients are: "a wild mixture of substances created by plants and animals for completely non-food purposes—their survival and reproduction." We eat them at our own risk.

This has never been a viable strategy, though. A smarter approach is to exploit people's biases, to create new products and market them as natural. Bottled water is the most successful example of this.

Now, there is an alternative to this essentialist theory, one that is often presented as an explanation of putatively irrational preferences and that has, I think, considerable merit. This is that bottled water is a sign of status. It is an example of what the sociologist Thorstein Veblen called "conspicuous consumption," a way to advertise how much money you have or, more generally, to show off your positive traits as a person. If the water were free or had obvious health benefits, it would be useless as such a signal, and, according to the signaling account, fewer people would drink it.

This signaling theory has considerable scope. It is often applied to the purchase of modern art. Any schmoe can buy, and appreciate, a pretty painting, while spending millions of dollars on abstract art might display a combination of wealth and discernment. Once you start thinking about signaling, you see it everywhere. I've sometimes wondered if signaling can explain why expensive private schools teach Latin. The schools insist that it is an intellectually worthwhile pursuit, but the alternative is that it's popular just because it hits the sweet spot of difficulty, association with power . . . and total uselessness, making it an ideal signal of status. If Latin turned out to help children learn other languages and improved their minds in certain ways, then public schools might start to teach it, and a proponent of signaling theory would predict that private schools would give it up and have their students spend an hour a day on Sanskrit or calligraphy.

This sort of theory is usually thought of as signaling to other people and, as a strategy, this is what it's for. Perhaps, though, we also signal to ourselves. I might want to reassure *myself* that I'm the sort of person who can afford, and who cares enough, to pay for something special, and so I might buy Perrier for my own private use. As the advertising jingle goes: Because I'm worth it.

Even if signaling plays some role, though, one still needs essentialism to explain other factors, such as fears of genetically modified foods, beliefs about cannibalism, and the use of food as aphrodisiacs. Essentialism explains our intuition that the invisible properties of what we eat, such as the courage of a warrior or the purity of bottled water, will pervade us. It cannot all be signaling, then; the evidence points as well to an essentialist mind-set.

TASTY

As the founder and CEO of Perrier North America, it was important for Bruce Nevins to convey to people how good his product tastes. It was a bad day for him, then, when he was on a live radio show and asked to pick out the Perrier from a selection of seven cups of water. He got it on the fifth try.

There is nothing wrong with his taste buds. In blind taste tests, with waters at equal temperatures, it is almost impossible to tell the difference between tap water and luxury bottled waters.

I would bet, though, that once Nevins left the radio show and went back to his life, he still thought that Perrier tasted really good—the radio test didn't prove otherwise. If so, he would be right. That is, someone who prefers the taste of Perrier to other waters but fails a blind taste test is not dishonest or confused. Perrier does taste great. It's just that to appreciate its great taste, you have to know that it is Perrier.

There have been several studies showing that how you think about food or drink affects how you judge it. The design of these studies is usually simple. You get two groups of people, you give them the very same thing to eat or drink, but present it to the groups in different ways. Then you ask how they like it. Studies find, for instance, that

- protein bars taste worse if they are described as "soy protein."
- orange juice tastes better if it is bright orange.
- yogurt and ice cream are more flavorful if described as "full fat" or "high fat."
- children think milk and apples taste better if they're taken out from McDonald's bags.
- Coke is rated higher when drunk from a cup with a brand logo.

This last study has been replicated with a brainy twist, where subjects were in an fMRI scanner. When given a blind taste test between Coke and Pepsi, with the liquids squirted into the subjects' mouths through a tube, the brain's reward system lights up and people are evenly split. But when they are told what they are drinking, a different pattern of brain activation emerges: people's preferences shift according to the brand they like more.

The findings that are most provocative have to do with wine. You can take the same wine and label it in different ways, and this affects how people, including experts, rate it. In one study, a Bordeaux was either labeled as a "grand cru classé" or as a "vin du table." Forty experts said the wine with the fancy label was worth drinking, while only 12 said this of the cheap label. The grand cru was "agreeable, woody, complex, balanced and rounded," while the vin du table was "weak, short, light, flat and faulty."

It gets worse. You might think that, at minimum, the difference between red wine and white wine would be obvious. But maybe not. At a party, bring out some white wine, put it in a black glass, and ask your friends what they think of the red wine you are giving them. When Frederic Brochette did this, many wine experts tasted it as red, and described it as such, using terms like "jamminess" and "crushed red fruit."

My favorite recent finding was reported in a working paper called "Can People Distinguish Pâté from Dog Food?" They can't. If you grind up a product called "Canned Turkey & Chicken Formula for Puppies/Active Dogs" in a food processor and garnish it with parsley, people cannot reliably distinguish it from duck liver mousse, pork liver pâté, liverwurst, or Spam.

THERE ARE two ways to make sense of what's going on here.

One is that there is a two-stage process. First, how you taste something is based on physical properties of what is tasted—it is in the nose and in the mouth. Then, as a second step, your belief about what it is that you are tasting transforms and modifies and elaborates the memory of the taste.

I witnessed a conversation between an adult and a four-year-old named Jonah, in which the preschooler was explicit about these two stages.

Adult: What do you like more, frozen yogurt or ice cream?

Jonah: They both taste just the same. I actually like frozen yogurt more.

Adult: Why do you like it more, if they both taste the same?

Jonah: Tasting it was the happiest moment of my life. I'm usually very very happy. When I tasted frozen yogurt at my grandma and grandpa's house, when I was tasting it, I was very very very very happy.

Jonah is making a distinction here between how something tastes and how much he likes it. Ice cream and frozen yogurt taste the same, but he likes frozen yogurt better. Maybe this is how knowledge can affect preferences. It doesn't change the experience itself but instead the value that we give to the experience, and this alters how we talk about it and think about it.

The second possibility is stronger—belief affects experience itself. That is, people don't say, "This tastes like a so-so wine, but since I know it's a grand cru, there must be more to it." They say, "Yum!"

The psychologist Leonard Lee and his colleagues did a clever experiment to distinguish these possibilities. They went to local pubs in Cambridge, Massachusetts, and asked people to taste "MIT brew"—Budweiser or Sam Adams with several drops of balsamic vinegar added. It turns out that in a blind taste test, people tended to prefer MIT brew to beer without vinegar—but if you just ask them, they *think* that vinegar makes beer taste worse.

The main experiment was done with a different group of subjects. Half of the subjects were first told that vinegar was added and then they drank the beer, and half first drank the beer and then were told that the vinegar was added. Both groups were then asked how much they liked MIT brew.

The logic is this. Suppose the weak theory is right—you taste what hits your tongue and what you know affects your opinion of that taste. If so, then it shouldn't matter when you hear that vinegar was added. If you think that it makes beer taste worse, then it should affect your perception of its taste. But if the strong option is right, timing should matter. If people are told that the beer has vinegar in it before they drink, they should taste it as worse, because this knowledge colors their experience. But if they are only told afterward, it's too late, they've already done the tasting, and so this knowledge can't affect the experience itself.

The strong theory wins. If you expect it to taste bad and then drink it, it tastes bad. But if you already tasted it, knowing about its status doesn't make a difference. At least for beer, expectations affect our experience itself, not our after-the-fact construal of the experience.

This conclusion is reinforced by a clever study that scanned people's brains while they tasted wine. It was always the same wine but it was described as costing either $10 or $90. As you would expect from the studies described above, people reported liking the wine more when it was described as expensive. What is more interesting is that while some parts of the brain were insensitive to the pricing manipulation (that is, at a brute sensory level, the brain was responding only to the sensations of taste and smell), the overall pattern is consistent with a fusion effect, in which the flavor expectations become integrated with the low-level sensory experience. This is proposed to occur in the medial orbitofrontal cortex, which is the same part of the brain that was activated in the Coke/Pepsi study described earlier.

A similar study was done in which the scientists presented people with an odor described as either "cheddar cheese" or "body odor" (it was isovaleric acid with some cheddar cheese flavor); this description had the expected effect on their experience and led to an activation difference in the same part of the brain. This is reminiscent of an episode of a television show I once saw (*Family Guy*), in which one character sniffs and remarks, "It's either bad meat or good cheese." The studies suggest that once you know the answer, you'll experience the smell differently.

I don't want to overstate the power of expectation. If taste were entirely a matter of what one believes, people wouldn't need taste buds and olfactory bulbs. These have evolved, after all, to provide us with information about the external world. We might not know much about a food, and have a bite to see whether we like it. Sometimes our physical experience can override our beliefs: "I know this is a vin du table, supposedly nothing special, but it is the best wine I have ever tasted," or "I know this flesh has the essence of a great warrior, but, ugh, it's still kind of rank."

The point, then, isn't that sensation plays no role in experience. It is rather that sensation is always colored by our beliefs, including our beliefs about essences. This can lead to a mutually reinforcing cycle. Suppose you think Perrier is purer than tap water, somehow superior. This enhances your experience of how it tastes: when you drink Perrier, you enjoy it more. This, in turn, reinforces your belief, which enhances your taste, and so on. If you believe that genetically modified foods taste odd, you will experience them as tasting odd, which will support your assumption that there is something wrong with genetically modified foods, which will make them taste worse in the future, and so on.

This sort of loop is not special to food and drink. If you are an audiophile and believe that expensive speakers significantly enhance your experience of music, then you will be biased to experience this, which will then reinforce your belief about the value of expensive speakers. It is not even special to pleasure. Suppose you believe that gay men are effeminate. This will affect your experience, and you'll be more prone to interpret an action by a gay man as being effeminate than if you saw the same behavior in a straight man. Your experience—hey, that gay man was quite effeminate!—will thus reinforce your stereotype. By distorting experience, beliefs, including essentialist beliefs, garner support for themselves, which is one reason why it is so hard to change our minds about anything.

PLEASURE, PAIN, AND PURITY

We might never know what it is like to be a dog or cat, but their behavior, physiology, adaptive niche, brain structure, and neurochemistry give every suggestion that they get pleasure from food. What's uniquely human, however, is our rich belief system about what we eat and why we eat it. It wouldn't matter to a dog, say, whether its food is natural versus artificial, made by a loved one or a despised enemy. Putting the word "Perrier" on the water bowl will not make the dog drink any faster.

There is also a difference between what people like and what people choose. For me, Coke tastes better than Diet Coke, but I drink Diet Coke because of the calories in Coke. Human choices can be dissociated from pleasure; not so for other creatures. If my dog goes on a diet, that's my choice, not hers.

Finally, there is a self-consciousness to our pleasures. Humans can observe the pleasure or pain that we experience, and can get further pleasure or pain from this observation. Emotions can feed on themselves. You can enjoy being with your friends, say, and thinking about your happiness might please you—you are a bon vivant, making the best of life, which is a pleasing thought. The flip side, more familiar to some of us, is that one can feel miserable about feeling miserable.

What is more interesting is that we get pain from pleasure and pleasure from pain. Only humans would enjoy this recipe from the (fortunately fictional) *Masochist's Cookbook*:

Cinnamon Spiced Pecans with Orange Rum Glaze

2½ cups raw pecans
1 cup rum
2 tsp. light-brown sugar
¼ tsp. salt
½ tsp. ground cinnamon
zest of 1 orange

Toast pecans at 350 degrees for 5 min. In a large saucepan, add rum, sugar, salt, cinnamon, and zest. Bring to a rolling boil. At this point, you may want to call 911. Remove pants. Bite down on an oven mitt and pour scalding glaze mixture over genitals. *Serves: 4*

This is an extreme form of masochism, involving serious bodily harm. A milder form is what Rozin and his colleagues have described as "benign masochism"—we seem to enjoy experiences that contain a bit of the nasty. Hot baths. Roller coaster rides. Pushing ourselves to the limit while running or lifting weights. Horror movies. It's not that we like them despite the pain; we like them, at least in part, *because* of the pain.

There are different theories why. Maybe it is the pleasure of the adrenaline rush. Maybe these are macho displays of how tough we are—more signaling. Perhaps there are opiates that get triggered along with pain, and the high from the opiates comes to surpass the low of the pain. I have my favorite theories, which I'll talk about in a later chapter, but here I just want to note, as Rozin does, that this happens all the time with food. Some very common foods and drinks are aversive. Few people enjoy, at first, coffee, beer, tobacco, or chili pepper.

Pleasure from pain is uniquely human. No other animal willingly eats such foods when there are alternatives. Philosophers have often looked for the defining feature of humans—language, rationality, culture, and so on. I'd stick with this: Man is the only animal that likes Tabasco sauce.

Then there is pain from pleasure. A mild version of this concerns violations of etiquette. Eating for a human is about more than sensory pleasure and biological necessity; it is a social act fraught with meaning. The rules of eating differ from culture to culture, but there are always rules—here you are supposed to burp, here you use a spoon, here you use your right hand but not your left. Violating these rules can lead to shame and guilt. Leon Kass, in his fascinating book *The Hungry Soul*, takes this further, suggesting that eating practices show a self-conscious recognition that we differ from other animals. For Kass, our response to violations, in ourselves and others, reflects a concern about our humanity.

Kass worries that these rituals are eroding, and in a sense he's right. Prohibitions about eating in public are just about extinct, and eating is often stripped of its social meaning. By one estimate, about one in five meals in America is eaten in the car, and, because of this, one of the major food inventions of the last century was a way to eat chicken one-handed—the chicken nugget.

But while etiquette might be fading, morality is expanding to take its place. Food is a particularly moral domain. There are some things that you shouldn't eat. Many people are morally appalled by the suffering of animals who are bred for our food. And recall that one objection to cannibalism, even consensual cannibalism, has to do with morality. You can doubt that Brandes, Meiwes's victim, was competent to make the choice that he did; and even when you eat someone who died naturally, this might be seen to show lack of respect, perhaps a general disregard for human dignity.

The philosopher Kwame Anthony Appiah has a revealing discussion about purity and politics, in which he observes that conservatives might well be obsessed with the morality of sex, but liberals have a similar obsession with food. As he puts it (admitting that this is a bit of a caricature), the liberal sophisticate

> *prizes organic foods that are uncontaminated by pesticides and additives, and shudders at how agribusiness has despoiled the environment. His commitment to organic, locally produced food is more than a consumer preference; it's a politics and an ethics.*

We are sometimes tempted to dichotomize our desires into simple animal appetites versus more civilized human tastes. But perhaps no such dichotomy exists. Even a pleasure such as the satisfaction of hunger is affected by concerns about essence and history, moral purity and moral defilement. There is always a depth to pleasure.

[illegible]

[illegible]

[illegible]

[illegible]

[illegible]

3

BEDTRICKS

IMAGINE DISCOVERING THAT YOU WERE WRONG ABOUT whom you just had sex with. Perhaps you thought he was your husband, but it was his twin brother. Or you had believed she was a prostitute, but it was your wife, disguising herself to test your fidelity. Perhaps the confusion or deception is not over who you are sleeping with, but over what—someone you had thought was a man was a woman, or a woman was a man, or an adult was a child, or a stranger was a relative—as with Oedipus, doomed to marry his mother and kill his father. In fiction, a person might find out that the sexual partner was a robot, monster, alien, angel, or god.

The term *bedtrick* was coined by Shakespearean scholars who were struck by the repeated appearance of this event in his plays. In her extraordinary book on the topic, the religious scholar Wendy Doniger points out that you cannot find a genre, place, or historical period in which the bedtrick isn't a repeated theme. We are obsessed with it—and always have been.

There is a lovely passage, for instance, in the *Odyssey*, written some 2,500 years ago, in which Odysseus returns from his travels, but his wife spurns his advances, unsure whether he is really her husband. Odysseus is angry, but Penelope insists, telling him that they have to sleep in separate rooms. She starts to arrange for their marriage bed to be moved out of the bedroom—but he points out that the bed can't be moved; he reminds her how he built it. She is now sure who he is, but by this point, he's furious with her for doubting him. She begs for his forgiveness:

But don't fault me, angry with me now because I failed,
At the first glimpse, to greet you, hold you, so . . .
In my heart of hearts I always cringed with fear
Some fraud might come, beguile me with his talk;
the world is full of the sort,
cunning ones who plot their own dark ends.

A bedtrick can be a fantasy, blameless infidelity in which you can sleep with someone new while still being true to your vows. More often, though, it is a nightmare. A bedtrick can be, legally and morally, rape—particularly humiliating in that the victim is tricked into complicity. Typically the victim is a woman; though a popular fictional variant is for a straight man to be fooled into sex with another man. The revelation can lead to revulsion; in *The Crying Game*, after discovering that Dil has a penis, Fergus becomes physically ill and vomits.

The Hebrew Bible is full of bedtricks. One of the best-known stories is Jacob fooling his father into believing that he is Esau and handing over his birthright (no sex, but it does happen in bed). When Lot's daughters get him drunk and have intercourse with him, it is a bedtrick of a sort; and a clear case is when Tamar disguises herself as a prostitute to have sex with her father-in-law. Most famously, there is the story in which Jacob works several years for Laban for the right to marry Rachel, but on the wedding night, Laban tricks Jacob by switching daughters: "And it came to pass, that in the morning, behold, it was Leah: and he said to Laban, What is this thou hast done unto me? did not I serve with thee for Rachel? wherefore then hast thou beguiled me?" (In symbolic recognition of this event, the contemporary Jewish wedding ceremony has the groom himself lowering the veil over the bride's face, so he can be reassured that he is marrying the right woman.)

The bedtrick nicely illustrates how sexual pleasure is not merely a matter of physical sensation. It is also rooted in beliefs about who someone really is and what someone really is. I will argue in this chapter that our essentialism can provide a new way to make sense of sex and love.

To tell this story, though, I have to start at the very beginning.

SIMPLE SEX

The simple story of pleasure is that animals evolve to enjoy what's good for them; pleasure is the carrot that drives them toward reproductively useful activities. (Pain is the stick.) It feels good to drink when thirsty and eat when hungry, because animals that were inclined to feel such joys left more offspring than those that weren't.

This logic easily applies to sex. If one animal seeks out opportunities for mating and the other is indifferent, then, all else being equal, the first will have more offspring in the future. From an evolutionary perspective, chastity is genetic suicide: You can't have offspring without sex, and sex, like food, is the sort of thing that one usually has to work to get; it won't just come to you. We have therefore evolved a motivation to seek it out, as have dogs, chimps, snakes, and many other creatures.

This appeal to natural selection is uncontroversial. There are all sorts of human activities whose adaptive value is unclear and that don't seem to relate in any obvious way to the activities of other species. We can reasonably debate the evolutionary origins of the pleasures we get from, say, music, the visual arts, or scientific discovery, and this is some of what we'll be doing in the rest of the book. And some aspects of sex are plenty mysterious. (Is female orgasm a biological adaptation or an anatomical accident? Why are some people exclusively homosexual? What is the origin of sexual fetishes?) But the pleasure of sexual intimacy poses no puzzle at all. Enjoying sex has a lot to do with having sex, and having sex has a lot to do with having children. It is hard to think of a better example of how a desire would be the outcome of natural selection.

But this simple evolutionary analysis doesn't get us very far. It tells us little about the precise nature of this evolved desire. Perhaps there isn't more to be said. One can imagine that we have evolved a "sex drive," some indiscriminate impulse toward rutting, and nothing more. One commentator describes the male toad as follows:

> *If a male sees something moving, there are three possibilities: if it is larger than I am, I run away from it, if it is smaller, I eat it, and if it is the same size, I mate with it. If the creature with which it is mating does not protest, it is probably the right species and the right sex.*

In her discussion of this, Wendy Doniger notes: "We all know men like this toad." Now, most would agree that human sexuality is more complicated than this, even for men, but perhaps this added complexity has little to do with evolution. We know, after all, that much of sexual activity is reproductively useless, such as masturbation, homosexuality, and intercourse with contraception, and so these specific activities could not have evolved through natural selection. Perhaps the human story of sex more generally will be best explained through personal history, cultural immersion, and free choice.

I have some sympathy for this view. For instance, feelings of sex and love have evolved to motivate our behavior toward real people, but, as I will discuss later, we can generate unreal people as the targets of sexual and romantic feelings. For humans, and for no other creatures, sex and love have moved from the real world into the world of the imagination. This is not an adaptation; it is an accident—a deeply significant one.

At the same time, though, the simple "sex drive" theory is too simple; our evolved predispositions turn out to be rich and structured. This becomes clear once we consider sex differences. While some tiny creatures have just one sex and reproduce by cloning, most fall into the categories of male and female. For reproduction to work, then, the sex drive has to be somewhat discriminate—after all, even the male toad is smart enough to guide itself toward female toads.

But there is more. Indeed, one of the victories of evolutionary biology is that it answers certain hard questions about sex differences. Why are animals with penises bigger and more violent, on average, than animals with vaginas? Why are the animals with vaginas typically choosier than those with penises, and why do the animals with penises often have attractive displays, such as the elaborate plumages of peacocks, or specialized weapons, such as the enormous tusks of male elephant seals?

These are the sorts of puzzles that flummoxed Darwin, but they are elegantly explained by the theory of parental investment, as developed by the evolutionary biologist Robert Trivers in the 1970s and refined extensively in the years that followed. The starting point is that our minds and bodies are adapted through natural selection for reproductive success, but there is typically a difference in the ideal strategies of males and females, one that reflects the asymmetry between sperm and egg. Sperm are tiny, multitudinous things, just barely genes and a motor to help them move toward the egg. Eggs are relatively enormous and contain all the machinery for growing a human. Further, in the standard mammalian plan, fertilization takes place inside the female and then, after birth, the baby is fed through the female's body. For a male mammal, then, the minimum investment required to create a baby, and thus pass on the genes, is a few moments of insertion and ejaculation. For a female it is months or years.

This makes a big difference, because while the female is growing and feeding a baby, she can't have another baby. As a result, one male can have children with many females at once, but not vice versa.

Trivers's insight was that this discrepancy entails a male-female difference in optimal reproductive strategy. Females should be prone to invest more in their offspring than males, because they can have fewer, and so each one matters more. This predicts that females should tend to be choosier when selecting mates, with an eye out for mates with the right genes and, in species in which this is an option, with the inclination to stick around and protect them and their offspring. Because males benefit from being chosen by females, there should be a corresponding trend for males to compete with one another for access to females, so they tend to be bigger and stronger, and often have evolved special weapons. They also advertise themselves to females, and so have evolved traits such as elaborate tails and markings. This is why the flamboyant tail belongs to the peacock, not the peahen.

This captures the usual differences between males and females, but what makes it such a convincing explanation is that it makes strong predictions about where sex differences should occur and where they shouldn't. From the standpoint of this theory, it is not the genitalia per se that matters; there's nothing magical about having a penis or a vagina. It is just that animals with penises tend to have less of an investment in their offspring than those with vaginas. This makes the neat prediction, then, that in the rare cases where the male-female investment is identical or flipped, the sex differences should change accordingly. And this is what happens. If you have a species in which parental investment is equal, either because the male and female work together to protect extremely fragile offspring (penguins) or because they just spray their sperm and eggs into the sea and offspring don't need more care after that (species of fish), you get physical and seductive equality of the sexes. If you have a species in which the males take care of the children, and the females are anonymous egg donors, you get choosy males and bigger, aggressive females with showy plumage.

• • • •

MORE COMPLICATED SEX

Where do humans fit in here? As the geographer and author Jared Diamond concludes in *Why Is Sex Fun?*, human sexuality is in most regards typical for species in which offspring are internally fertilized and benefit from care from both parents. We are not adorable pair-bonding penguins, but we are also not lions, wolves, or chimps, animals where the males don't even know who their offspring are. (Scientists who do care need to do DNA testing to figure out which child belongs to which male.) We are in between.

Our evolutionary history is reflected in our bodies. Size differentials between males and females within a species reflect the extent of competition for mates among males, which in turn relates to the difference in parental investment. This is why it's hard to tell a male penguin from a female penguin—they are egalitarian co-parents. Male humans are, on average, quite a bit bigger than female humans—we are not penguins—but the male-female difference is not as large as in species in which males have nothing to do with children.

Our evolutionary history is also reflected in our minds. Human males tend to have more of an interest in sex with multiple partners and are more easily aroused by, and interested in, anonymous sex. As far as we know, this is true everywhere on Earth; the study of sex differences is one of the few areas in psychology in which scientists have done the relevant cross-cultural research. Prostitution exists largely to satisfy this male desire for variety, as does pornography. There are male prostitutes and depictions of male nudity and male sexuality in pornography, but, for the most part, these exist for gay men.

This begins to sound like the story—attributed to Dorothy Parker and to William James, among others—of the writer who wakes up in the middle of the night convinced she has made a great discovery, writes it down, goes back to sleep, and wakes up to discover that she has written this:

Hogamous Higamous
Man is polygamous.
Higamous hogamous
Woman monogamous.

Statistically, it is on the right track, but it is incomplete. We need to explain the fact that men are often monogamous, and women are often polygamous.

One consideration is that human children are particularly fragile creatures, born far too early, with a long period of dependence on adults for food and shelter and protection from predators both animal and human. Fathers matter, then, as they help in protecting and raising the children, and also because they protect the mother (because if she dies when the baby is feeding, the baby is likely to die too).

This does not mean that mothers and fathers are interchangeable. There remains an evolutionary battle of the sexes, because it is in the male's genetic interest to fool around on the side. This would be bad news for the female, who would be better off with a mate who sticks with her and her child instead of distributing his time and resources to other offspring and women. This conflict shapes female preferences about whom to mate with—they are looking for males who show signs of future fidelity. Men might evolve to fake these signs, but if women are good at seeing through this deception then males who tend to be sexually and romantically faithful might out-reproduce the cads. It is attractive to be faithful. In this way, sexual selection will serve to narrow the gap between men and women's sexual preferences.

There is a further wrinkle in this. Human females have the strikingly unusual feature of hidden ovulation. They can have, and enjoy, sexual intercourse anytime during the menstrual cycle. One theory of this is that when there is overt ovulation—the mammalian status quo—it is easier for males to fool around while still ensuring that their children are their own. They just need to monitor their own mate during specific times to make sure that she doesn't have intercourse with another male, and can spend the rest of their own time looking for females who are unattached or who have inattentive mates. But if human females can mate all the time, and if it is unpredictable when this mating will lead to a child, it forces the male to stick around. If not, he runs the risk of wasting resources on a genetically unrelated child.

(As an aside, the logic here assumes that female infidelity is a fact of our evolutionary history. This sort of ovulatory blackmail only works if women sometimes fool around. Female infidelity plainly exists in the here and now, including infidelity of the genetically relevant type—some men are unknowingly raising children whom they are biologically unrelated to. And there is physical evidence that female infidelity existed over the evolutionary time span: the big testicles of human males relative to other primates. This is consistent with a "sperm wars" account in which females mate with multiple men, making it adaptive for men to increase sperm production. So the *Higamous hogamous* stanza is not quite right either.)

We have considered how the helplessness of children provides an evolutionary basis for monogamy, but there is another consideration, something special to human relationships. We can be smart and we can be kind. Smart enough and kind enough, for instance, to divert ourselves through fantasy, to deny ourselves pleasures that we believe to be wrong, to take the perspective of another person, to rationally compute costs and benefits, and so on. We can choose to be like those adorable penguins.

HEADTRICKS

Here is something that men and women have in common: We all like to look at pretty faces.

This is not just sexual. Straight men and women enjoy looking at attractive same-sex faces. Regardless of their sex, good-looking faces light up the brain, triggering neural circuits devoted to pleasure. Even babies, who (Freud aside) have no sexual urges at all, are suckers for a pretty face and prefer to look at one from the very start.

This baby result would have surprised Darwin, who believed that standards of beauty are culturally arbitrary and so have to be learned. But there are features that everyone everywhere finds appealing: Unblemished skin. Symmetry. Clear eyes. Intact teeth. Luxuriant hair. Averageness. This last one might seem surprising, but if one picks out 10 faces at random, either 10 men or 10 women, and morphs them together, the result would be good looking, and when shown this composite face, babies would probably rather look at it than at any one of the individuals. So would you.

Why would these considerations matter? Factors such as smoothness of complexion, symmetry, clear eyes, intact teeth, and good hair are overt cues of health and youth, which are good things for everyone to attend to when looking for a mate. This is particularly the case for symmetry; it is hard to be symmetrical, and bad things such as poor nutrition, parasites, and simply the ravages of time eat away at it. It is a mark of success.

It is less clear why averageness is good. It might be that it reflects health, on the logic that most deviations from normal are bad. Averageness also corresponds to heterozygosity, or genetic diversity, which is another good thing. A very different possibility is that average faces are in a literal sense easy on the eyes; they require less visual processing than nonaverage faces and we tend to prefer visual images that are easier to process. One wrinkle here is that while average faces look good, they don't look *terrific*—the most attractive faces are not the average ones. (When you do these morphs, you get a fine face, but not one with movie-star good looks.) Perhaps it isn't that average faces are positively attractive; it's that nonaverage faces run more of a risk of being unattractive.

I have always found it odd that there are no big sex differences in our judgments of attractiveness. Looks do matter more to men than women, not just in America and Europe, but everywhere else in the world that the question has been asked. But, with just one exception, there is no difference in what the sexes see as attractive. Straight men are just as good at appreciating a handsome male face as straight women are.

The exception is that women's preferences shift during the menstrual cycle. Most of the time, they are drawn to male faces of the sort that meet the criteria above, but when they are ovulating, they are also drawn to highly masculine, chiseled faces. When I first heard of this finding, it seemed too cool to be true, but it has been replicated now several times. One interpretation is that, when ovulating, women are looking especially for good genes and have an eye out for hypermale males.

IN AN INTRIGUING series of studies in the 1950s, investigators were interested in what features would initiate sexual behavior in male turkeys. They first found that you could get arousal with a lifelike model of a female turkey—the males would gobble, strut, puff up, and eventually mount the model. To find the minimal stimulus for sexual response, the scientists removed parts from the model, such as its tail, feet, and wings, ultimately ending up with a head on a stick. The males were fully aroused by this head, and would prefer it even to a headless body.

People can be turkeys. We are hardwired to be attracted to certain perceptual cues, and this can be triggered without a real person to go with them, as when we are sexually aroused by two-dimensional arrays of pixels on a computer screen. Even when we are with actual individuals, we can be transfixed by a body part while being indifferent to the person who comes with it.

This shows up with fetishes, in which sexual arousal can become focused on a specific body part. An extreme example is the serial killer and foot fetishist Jerome Brudos who started off by stalking women, choking them unconscious, and running away with their shoes. He then progressed to rape and murder, keeping one woman's foot as a trophy. Then there is the foot fetishist described by the writer Daniel Bergner—a kind and romantic man who was tormented with powerful involuntary lust. He would become aroused by a surprise glimpse of exposed female feet in the summer, and he tried not to listen to the weather report in the winter because of the painful erotic distraction caused by phrases like "a foot of snow."

Sometimes, then, sexual desire can be triggered in simple ways. Being smart creatures, we can work to appeal to other people on this perceptual level. One doesn't need a PhD in cognitive ethology to cover up that zit. People work hard to modify their faces; mostly they try to look younger, with lipstick, blush, eyebrow plucking, wigs, toupees, hair implants, and so on. Plastic surgery and the neurotoxin botox are also used, along with low-tech methods such as pinching one's cheeks to make them look red, an ancient trick. Some techniques extend below the neck, of course, such as muscle building, breast implants, and penis enlargement.

Just as people can consciously try to fake cues that elicit sexual interest in others, such as by putting on makeup, we can also see through the fakery of others. As essentialists, we want the real thing. Most women, for instance, would prefer a man posessing what they believed to be naturally strong features and youthful good looks rather than someone who bought this appearance through botox, hair plugs, and testosterone injections.

HOW IMPORTANT are looks? Even the most cynical evolutionary psychologist would concede that other considerations can override these innately primed cues of attractiveness. Women's choices are particularly influenced by factors such as wealth and status—a woman might choose the pudgy old millionaire over the hot young bodybuilder. But still, the cynical argument goes, our sexual and aesthetic responses are triggered by certain perceptual features. Clear skin trumps blemishes, symmetry is better than asymmetry, and so on. You might love your aging spouse more than the supermodel, but the supermodel will always be your dream date.

I disagree. Looks aren't everything when it comes to desire. The logic of adaptation says that we are attracted to those who have certain relevant traits—and some of these are not visible on the face or body. It is easy to be misled here by the research, because so much of it focuses exclusively on appearance, as with experiments that look at *Playboy* centerfolds and see what physical qualities they have in common, or that present photographs to undergraduates and get them to rank them, or that show computer-generated faces to babies and see which ones they prefer to look at. Such studies can tell us interesting things about what we find perceptually attractive, such as the importance of symmetry or averageness. But they are incapable of telling us about anything that cannot be captured in a picture. The same point applies to those clever studies in which people are asked to sniff the sweaty T-shirts of strangers. These tell us a lot about how pheromones affect sexual interest, but nothing about how important smell is compared to other qualities.

What else might matter? One consideration is familiarity. In one study, researchers got a team of women to attend different classes at the University of Pittsburgh. These women never spoke during the lectures and never interacted with the students. But the number of classes they attended varied—15, 10, 5, or none. At the end of the course, students were shown pictures of the women and asked what they thought of them. The women judged as most attractive were those who had attended class 15 times; judged least attractive were those the students had never seen before. This is a small study, but it fits a voluminous literature in social psychology on the "mere exposure" effect—people like what they are familiar with, which is a rational way for the mind to work given that, other things being equal, something you are familiar with is likely to be safe. Mere exposure applies to attractiveness, then, explaining some of the appeal of the girl (or boy) next door.

In another study, experimenters had people rate the photographs of classmates in their high school yearbooks for how much they liked them and how attractive they felt that they were. Strangers of the same age also ranked the photographs for attractiveness. If liking had been irrelevant, the ratings by classmates and the ratings by the strangers should have matched—but they didn't. The classmates' attractiveness ratings were swayed by how much they liked each person, further evidence that there is more to being good looking than looking good.

Even when you rate the faces of strangers, looks aren't everything. One study found that a main factor in attractiveness has nothing to do with averageness, symmetry, sexual dimorphism or anything like that—it's whether the person is smiling.

THREE QUESTIONS TO ASK WHEN YOU ARE LOOKING FOR A MATE

What else determines our sexual and romantic response to another person? There are three questions anyone looking for a mate needs to answer. I think these are interesting in their own right, but they also begin to give us some appreciation of the richness and complexity of human attraction.

1. Is the person male or female?

Freud claimed "when you meet a human being, the first distinction you make is 'male or female' and you are accustomed to making the distinction with unhesitating certainty." This is true for me at least: I get e-mail from strangers with foreign names and when I can't tell whether the sender is a man or a woman, it is oddly unsettling. It shouldn't matter—I have no intention of mating with them—but it does. When we see a baby in a diaper, the first question that many of us ask is: Is this a boy or is it a girl?

Maybe the baby is looking back and asking the same thing. In the first year of life, babies can distinguish male from female faces and they know that a male voice goes with a male face and a female voice with a female one. They like to look at females more, though it's not clear whether this is because of an innate expectation for female caregivers or because most children have female caregivers and they, like us, prefer what they are used to.

Babies become children, and come to form views about males and females. There are, of course, all sorts of differences between men and women, including psychological differences, most obviously in preferences about whom to have sex with, and social differences, such as who tends to be a nurse or police officer. Children quickly learn about these—for instance, both boys and girls know that girls tend to find feminine toys more interesting than boys will. This is hardly surprising. Such generalizations are true in the environments children live in and children are good at noticing true things.

What's more interesting is that children have theories about why these differences exist. The psychologist Marjorie Taylor explored this with an experiment in which she told children about a baby boy who grew up on an island that only had girls and women on it, and about a baby girl who grew up on an island with only boys and men. How would this affect the child? Would the boy, for instance, like to play with dolls? If you believe that this sort of behavior is the product of the culture, then yes; if it's intrinsic, then no. Taylor found that the children she tested tended to focus more on the innate potential: regardless of the environment, boys would do boy things, and girls would do girl things. It was only the adults who reasoned about socialization. This fits with interview studies finding that children start off with a biological orientation about how males and females differ. As mentioned in the introduction, children say things such as "Boys have different things in their innards than girls." Over time, some move to more of a sociological and psychological orientation—"because it's the way we have been brought up"—and presumably this is what they learn from our culture. Society makes us less essentialist, not more.

It is not just that we believe that males have some traits and females have others; we often believe that this is the way it *should* be. In Deuteronomy, it is a severe violation for a woman to wear a man's clothes, and vice versa, and many societies have laws forbidding women to participate in traditionally male activities, such as driving a car or joining the military. Even in liberal societies in which being homosexual or transsexual is not a crime, these acts are repulsive and immoral to many, and sometimes inspire violent reprisals.

Children in the United States often disapprove of sex-role transgressions, particularly those done by boys, such as wearing a dress. Some four-year-olds say that they would not want to be friends with such a person, that such behavior is wrong, and that they would be surprised and disgusted to see such behavior. Some even said that they would respond with violence if they saw such a thing. Children are not just sensitive to the boundaries between these categories, then; they are willing to police these boundaries.

2. Is the person a relative?

The psychologist Jonathan Haidt describes the following moral dilemma:

> *Julie and Mark are brother and sister. They are traveling together in France on summer vacation from college. One night they are staying alone in a cabin near the beach. They decide that it would be interesting and fun if they tried making love. At the very least, it would be a new experience for each of them. Julie was already taking birth control pills, but Mark uses a condom too, just to be safe. They both enjoy making love, but they decide never to do it again. They keep that night as a special secret, which makes them feel even closer to each other. What do you think about that? Was it ok for them to make love?*

I present this dilemma whenever I teach Introduction to Psychology, and the reaction I always get is: *Gross!* We are repelled by this, and most believe it to be immoral. But why? Indeed, why aren't we sexually attracted to our siblings? Many people have brothers and sisters that they would acknowledge as very attractive, but it is very rare that they want to have sex together. Few, if any, parents have to worry that their adolescent kids are sneaking off into the backseats of cars with each other, or booking romantic getaways with one another. Blocking brother-sister incest is not a big part of the educational program, ministers and politicians don't voice their disapproval, and psychologists don't get government grants on how to combat it. It is like eating feces; it's not a problem because almost nobody wants to do it.

Some incest prohibitions are often violated, but these concern the relationship between more distant relatives, where we don't feel the same sort of repugnance. Often these less intuitive prohibitions are written into law and scripture. Leviticus, for instance, is stern about sex between a man and his son's daughter or daughter's daughter, putting the rationale in language that is a poetic version of Richard Dawkins's selfish-gene theory: "you shall not lay bare her nakedness, for it is your nakedness."

The evolutionary rationale for incest avoidance is that it is a bad idea, genetically, to have children with your kin, because they share too many of your genes. This is known as "inbreeding depression"—the risk that recessive genes become more likely to be homozygous. But it is a lot harder to explain how this works in practice. Suppose we have a little book in our heads. In it is written, "No sex with close kin!" along with the emotional statement "It is really gross!" and the evaluative statement "It is morally wrong." Suppose every human chooses to follow its instructions. Still, the question remains: How do you figure out who your close kin are?

We can get a clue to the answer by considering cases where things go wrong. People sometimes do have sex with kin by mistake, as with the fictional Oedipus, the characters in the John Sayles movie *Lone Star*, or a real-world case in England in 2008, in which twins were separated at birth, met later in life, and got married. Then there are cases in which people who are not genetic kin think of one another as such. The most-studied examples of this are Israeli children who are raised together on a kibbutz, and arranged marriages in China and Taiwan where parents adopt a female baby to raise in their family and to later marry their son. In both cases, subsequent sexual and romantic relationships don't tend to occur.

These examples suggest that there is something about being raised together that kills the libido. In an important article in *Nature* in 2007, Debra Lieberman, John Tooby, and Leda Cosmides explored two ideas about what, precisely, that something is. The anthropologist Edward Westermarck thought up one possibility in 1891. It is duration of co-residence that matters: children raised close to one another for a long period of time will develop a sexual aversion later in life. A second potential factor is the observed association between a person's mother and another infant. If I see my mom breast-feeding some child, it's very likely that this child is a relative. As the authors note, this second cue can only be used by older siblings toward younger ones; the younger sibling has never seen the older one as an infant.

To compare these theories, they asked adults a series of questions—about whether they were raised with their siblings, how much they care about them, and how disgusted they are (if at all) at the thought of having sex with them.

They found that if the adult had no opportunity to see the sibling being cared for as an infant, then duration of co-residence is the big factor—the longer you live with the sibling, the more sexual aversion and more caring. But when people do witness their sibling being cared for as a baby, this trumps co-residence; you get high sexual aversion (and high level of caring) and duration of co-residence doesn't matter anymore. To put it differently, seeing someone interact as a baby with your mom is a libido killer later in life, even if you don't live with them that long.

This is all unconscious. You might know perfectly well that someone isn't biologically related to you, as in the kibbutz case, but if you spend your childhood with him or her, the natural reaction to incest kicks in. Conversely, you might know perfectly well that someone is close genetic kin, but if you have never lived with that person, the idea of sex doesn't gross you out. My guess is that this is the case for the British twins; once they found out that they were siblings, they were appalled, shocked, and so on (and they did have their marriage annulled)—but it didn't make their sexual and romantic feelings disappear.

IT IS NOT just siblings you have to worry about. It is desperately important that you know who your children are. You want to avoid them as sexual partners, and you want to devote your love and care to them.

For women, this is easy; women's children are the fruit of their wombs. It is men who have to worry. They can never be certain which children share their genes, and, as we now know from DNA testing, even when they think they know, they are often wrong. Many men are literally *cuckolds*, unknowingly raising the children of other men. The term comes from "cuckoo," which is a bird that lays her eggs in the nests of other birds—a nesttrick.

It seems likely that the cue mentioned above regarding sibling incest—duration of co-residence—applies to parent-child incest as well. If you live with your children from when they are babies onward, you probably don't want to have sex with them. The problem with stepfathers is that they often enter the family later. If a man has not been with his children when they were young, then he is more likely to be sexually attracted to them as well as more likely to be violent toward them (indeed, more likely to kill them).

Again, our gut feelings are driven by these cues, not by explicit knowledge. If you adopt a baby, your attachment will be just as strong as for a nonadopted child. You think about the baby as if he or she were your flesh and blood. On the flip side, a man who first meets his teenage daughter late in life might still feel attracted to her, even if a DNA test reveals that they are blood relatives.

A different potential cue to kinship is what the baby looks like. The more the baby looks like a given man, the more likely it is to be his progeny. This suggest that fathers will attend to their children's appearance to determine paternity, which has led some researchers to predict that babies would resemble their fathers more than their mothers—the idea is that they would benefit from signaling their genetic relatedness to the adult male around them.

It is not clear, though, that this prediction makes sense. If cuckoldry is common (and it has to be somewhat common if men need reassurance), this would be a *terrible* evolutionary strategy. Babies that are not the child of the adult male run the risk of being rejected or killed; they would look like the wrong guy. Indeed, while there was an initial study suggesting that babies look more like Dad, no other study has replicated this finding.

3. What is the person's sexual history?

Virginity has mattered as far back as we can tell. It is introduced in Genesis in its description of Rebecca ("And the damsel was very fair to look at, a virgin") and is mentioned repeatedly throughout the Hebrew Bible (the word is used 700 times, on one count). It is less of an ongoing theme in the New Testament, but of course virginity is at the very core of Christian belief, with the virgin birth of Christ.

Virginity, in this context, refers to not having had penetrative sex. This emphasis on penetration has mystified some people. In September 2007, the online magazine *Slate* asked the best-known sex columnists about what most puzzles them, and one entry—from Emma Taylor and Lorelei Sharkey (Em & Lo)—was on this topic.

> *We've never been able to understand why virginity is still defined strictly in terms of penile penetration. . . . how is it possible that a straight couple can engage in oral sex, manual sex, mutual masturbation, and possibly even anal sex (if you believe the rumors about Catholic school girls) and still claim they're "saving themselves for marriage"? Sure, intercourse's role in baby-making elevates it a bit among sexual acts. But these days, birth control, family planning, and reproductive technologies mean that intercourse is less a means to an end and more a pleasurable end in itself. Add to that the influence of feminism and the gay rights movement, and you'd think that there'd be a few more seats at the official sex table.*

It does seem arbitrary. It brings to mind the pointless debate in the late 1990s that was spawned by President Clinton's insistence that oral sex did not count as "sexual relations." But it is no mystery why we are biased in this way. Taylor and Sharkey answer their own question when they concede that the baby-making power of penetrative sex "elevates it a bit" among sexual acts. More than a bit!

True, there is now a separation between intercourse and babies. One can have sex without wanting to have babies and can explicitly take steps not to have them. Less frequently, there can be the begetting of children without sexual intercourse. But our minds, and our sexuality, are not rationally calibrated to modern times. We don't fully live in the here and now. Our desires have two histories, a personal one and an evolutionary one, and for most of the life of our species, penetrative sex was the only way to have a baby. It is not surprising that we give it a special status, different in kind from mutual masturbation, phone sex, and backrubs.

The central notion of virginity is even narrower than Taylor and Sharkey would have it. It is characteristically restricted to females (the English word "virgin" is derived from the Latin, meaning "young woman"). Female virginity matters more than male virginity because females are virtually always certain of who their children are, while men are often in doubt. It is an evolutionary disaster for a man to raise a genetically unrelated child, and so it matters hugely to him whom his partner has had sex with in her immediate past, with the best answer being: nobody.

The appeal of the virgin has led to the formation of some unusual markets, which reached a modern extreme with a 22-year-old women's studies student named Natalie Dylan, who is auctioning off her virginity on the Web. (She promises to ensure her chastity through a gynecological exam and a lie detector test.) Dylan is not the first to do this, but her auction was picked up by the national and international press, and she is getting offers of over a million dollars. Then there is sham virginity; in the United States, some married women pay for hymen reattachment, so that, as a gift to their husbands, they can simulate being virgins.

The obsession with virginity is one of the ugliest aspects of our sexual psyche. In many societies, there are rituals of virginity testing before marriage, and various forms of genital mutilation to enforce chastity by making it difficult and unpleasant for a woman to have penetrative sex. There are acts of terrible violence against women who are found to be unchaste, including against those who have been raped. The obsession with virginity motivates the sexual exploitation of young women and girls, and, due to a horrible extrapolation of the idea of purity, has motivated the myth that sex with a virgin is a cure for AIDS.

DEEPER

Even when you narrow down the candidates to those of the right sex, relation, and history, still, it is hard to choose a long-term mate. When he was 29, Charles Darwin agonized about whether to marry. In 1838, he wrote down the pros and cons. These are shown on the opposite page. He then wrote "Marry-Mary-Marry Q.E.D." and, months later, that's just what he did.

Darwin's pros and cons are a nice mix of the Victorian and, well, the Darwinian. Children are at the top of the "Marry" list, but they also make it onto the "Not Marry" one because of the expense and anxiety. Sex isn't explicitly mentioned, though physical contact is discussed. But the main theme of the pro side is not sex or kids. It is the view that marriage would enrich Darwin's life, providing him with a friend and companion.

In one of his love letters to Emma Wedgwood, a week before they married, Darwin wrote, "I think you will humanize me, and soon teach me there is greater happiness, than building theories & accumulating facts in silence & solitude." She did; they had an extraordinarily close relationship, one that ended up affecting his work in a substantive way, as his concerns and respect for Emma's

Marry

Children—(if it Please God)—Constant companion, (& friend in old age) who will feel interested in one,—object to be beloved & played with.—better than a dog anyhow.—Home, & someone to take care of house—Charms of music & female chit-chat.—These things good for one's health—*but terrible loss of time.*—

My God, it is intolerable to think of spending one's whole life, like a neuter bee, working, working & nothing after all.—No, no won't do.—Imagine living all one's day solitarily in smoky dirty London House.—Only picture to yourself a nice soft wife on a sofa with good fire, & books & music perhaps—Compare this vision with the dingy reality of Grt. Marlbro' St.

Not Marry

Freedom to go where one liked—choice of Society & *little of it.*—Conversation of clever men at clubs—Not forced to visit relatives, & to bend in every trifle.—to have the expense & anxiety of children—perhaps quarelling—**Loss of time**.—cannot read in the Evenings—fatness & idleness—Anxiety & responsibility—less money for books &c.—if many children forced to gain one's bread—(But then it is very bad for one's health to work too much)

Perhaps my wife won't like London; then the sentence is banishment & degradation into indolent, idle fool—

religious views tempered his claims about how evolution has shaped the human mind.

When looking for a mate, Darwin was looking for more than bilateral symmetry and the right hip-waist ratio. He wanted a good and special person. You can read youth and health from the face and body, but one also looks for qualities like intelligence and kindness. Smart and kind people do well in the world, and so do their children. You also want someone who will faithfully take care of the children, and someone who will help and support you. It is not surprising that in the largest study ever of human mate preferences, looking at people in 37 cultures, the most important factor for both men and women is kindness.

LIKE DARWIN, all of us are on the lookout for partners who are smart and faithful and kind. The problem is figuring out who they are.

This brings us to what biologists describe as sexual selection. Consider the flamboyant tails of peacocks. These are worse than useless—unwieldy and heavy, slowing the bird down, hard to keep clean, a "Kick Me" sign for predators. Before developing the theory of sexual selection, Darwin wrote that the sight of the peacock feather made him sick—it was a humiliating refutation of the logic of natural selection.

The solution that he arrived at is that these tails don't directly help with survival; they don't avoid predators or kill prey or provide warmth or anything else that helps the peacock better deal with the physical world. But they are appealing to peahens. If peahens prefer to mate with peacocks with a little bit of color, then the next generation will include both more colorful males and peahens with a similar taste in flamboyance, and then, over the course of evolutionary history, you end up with the peacock's tail.

In 1958, the evolutionary biologist John Maynard Smith extended this analysis to the complicated dances of male fruit flies. Such dances look useless, and they are—except when you take into account sexual selection. Females use these dances to decide whom to mate with, a reasonable evolutionary choice since you need to be fit to dance well. Picky females get their fit children, and the genes that motivate males to dance and females to look for dancers get spread throughout the population.

The psychologist Geoffrey Miller has argued that many of the more interesting and ostentatious aspects of human nature have evolved through sexual selection, as a way for people to advertise their worthiness to one another. They are ways in which we reveal our fitness, and Miller would include dance here, and much of sports, art, charitable activities, and humor. For him, the brain is a "magnificent sexual ornament."

I am not going to discuss Miller's grand theory in detail here, but there are two insights that he has about sexual attraction that are worth exploring. The first is costly signaling, which was mentioned in the last chapter's discussion of why people pay so much for bottled water. The idea is that displays of personal quality are only taken seriously if they involve some cost, some level of difficulty or sacrifice. If anyone can easily do the display, then it is worthless, because it is trivially easy to fake. Costly signaling shows up in the gifts we give to one another, particularly during courtship. Miller asks, rhetorically, "Why should a man give a woman a useless diamond engagement ring, when he could buy her a nice big potato, which she could at least eat?" His answer is that the expense and uselessness of the gift is its very point. A diamond is understood as a sign of love in a way that a potato isn't, because most people would only give one to someone they care about, and so the giving signals some combination of wealth and commitment.

Financial value is not the only signal of commitment. The economist Tyler Cowen points out that the best gifts for someone you live with are those that you, yourself, wouldn't want. He points out that even if his wife would enjoy the complete DVD set of *Battlestar Galactica*, it would be a lousy gift, because he would also get pleasure from it, and so the giving doesn't signal any particular love for her.

Other signals include changing your name, moving, and getting a large tattoo with your lover's name on it (and it can't be one of those stick-on tattoos that you rub off with hot water!). Marriage is obviously a commitment, and it becomes more costly (and more of a sign of love) if it is difficult to get divorced. Prenuptial agreements, however rational they might be, have the opposite effect, as you are explicitly signaling your worry that the relationship might end and shielding yourself from the costs. A man getting a vasectomy after his wife is no longer fertile is signaling that he won't leave her and have children with a younger woman (but, again, if the vasectomy is reversible, it's not as romantic.)

These are all signs of commitment, of love, though it should go without saying that this sort of costly signaling is not always welcome. Cutting off one's ear, for instance, is typically excessive, as is tattooing or self-mutilation after a first date. While these successfully signal interest and devotion, they also convey desperation and madness.

Miller's second neat idea is that of a "hot chooser." The idea is that when we choose mates, we are looking for people who give us pleasure. This might seem obvious at a personal level, but Miller explores it from an adaptationist perspective, as a force for the evolution of certain traits.

A simple physiological example is the penis. There are all sorts of oddities about the human body relative to other primates—males grow beards, females have enlarged breasts and buttocks and narrow waists—but the most striking difference has to do with the male genitals. Some primates have genitals that are more visually interesting than the human one. The mandrill has a bright purple-pink scrotum and red penis, vervets have a blue scrotum and red penis, and so on. But the human penis has a clear tactile advantage, being longer, thicker, and more flexible—very different from the small, pencil-thin penises of other primates, which are about two to three inches long and made rigid by a penis bone. Miller makes the controversial claim that this is the product of female sexual selection; females were drawn to males who gave them sexual pleasure, leading to the evolution of a better penis.

The brain, for Miller, has evolved much like the penis. People are on the lookout for entertaining mates. We prefer to be with, and mate with, those who make us happy. This puts evolution in a new light. Evolutionary psychologists typically see the mind as either a scientific data-cruncher, constructing theories of the natural environment, or as a Machiavellian schemer, trying to outfox others in a zero-sum game of social dominance. Maybe the mind is also an entertainment center, shaped by the forces of sexual selection to give pleasure to others, to possess the capacity for storytelling, charm, and humor.

• • • •

TRUE LOVE

The argument so far is that sexual desire can be smart. While we've evolved to be sensitive to the shape of the face and the curve of the hips, we also look at deeper factors, including sexual history, signs of commitment, and wit, warmth, and kindness.

Here I want to emphasize a further aspect of this depth, which is that we are not exclusively attracted to faces or bodies, or even to personality or intelligence. We are attracted to specific people who so happen to have these certain properties. We fall in love, after all, with individuals, not with aspects of people. As George Bernard Shaw put it, "Love is a gross exaggeration of the difference between one person and everybody else."

There are two reasons why love works this way. The first is its seductive power. If you stick with me for my intelligence, wealth, or beauty—as opposed to for me, myself—then our relationship is fragile. The psychologist Steven Pinker outlines the worry here:

> *How can you be so sure that a prospective partner won't leave the minute it is rational to do so—say, when a 10-out-of-10 moves in next door. One answer is, don't accept a partner who wanted you for rational reasons to begin with; look for a partner who is committed to staying with you because you are you.*

This commitment might seem irrational, but it is an attractive irrationality, and if the person is interested in you as well, this can be very attractive. "Murmuring that your lover's looks, earning powers and IQ meet your minimal standards would probably kill the romantic mood," Pinker notes. "The way to a person's heart is to declare the opposite—that you're in love because you can't help it." Indeed, neuroscientists have discovered dedicated systems for romantic love and for attachment, and some have argued that you can become addicted to a specific person the same way that one becomes addicted to cocaine—though the sort of addiction explored here is not romantic love but rather the love of a mother for her child.

The focus on individuals is not just a seductive strategy, though. The second reason why we fall in love with individuals is that we focus on individuals for *everything* that is valuable to us. This is how we reason about artwork, consumer products, and sentimental objects. If I owned a painting by Chagall, I would not be pleased if someone switched it with a duplicate, even if I couldn't tell the difference. I want *that painting*, not merely something that looks just like it. A knockoff Rolex is going to be worth less than a real one, regardless of how good it is, and when we replace children's security blankets or teddy bears with duplicates (something that we did in the laboratory—see the next chapter), they are not pleased.

As an illustration of this fact about love, think about the person you love the most. Now imagine that there is someone else in the world, someone who looks virtually identical to your special someone, so much so that most people cannot tell the two apart. Indeed, imagine that he or she is a genetic clone of your partner and has been raised in the same house by the very same parents.

In other words, imagine that your partner has an identical twin. If you were attracted to the properties of a person, rather than the person him- or herself, then your attraction should extend to a considerable degree to the twin. Interestingly, studies of people who are married to twins find that this doesn't happen. The romantic attraction is to the person you're married to, not his or her superficial qualities.

Sexual desire is similarly calibrated to individuals, not properties—though here it can be the *less* familiar individual who elicits the greater response. This is nicely illustrated in a play written by Isaac Bashevis Singer, one involving an accidental bedtrick. Singer tells of a fool who wanders away from his village of Chelm, gets lost, and ends up back at the village, except that, confused, he believes that he has come across another village in which people look identical to those where he came from. He sees his wife, whom he had been long tired of, and is powerfully aroused. At the perceptual level, she is of course familiar—but we are not perceptual creatures. As far as I know the experiment has never been done, but I would bet that the spouses of identical twins would be unusually affected, perhaps aroused, by the sight of their wife's or husband's naked twin, even though, at a perceptual level, the body itself is entirely familiar.

Indeed, variants of this experiment are being done online each day. Porn sites boast about pictures of naked celebrities captured from movie clips or, in some cases, from telephoto lenses. What presumably makes these pictures arousing isn't the visual experience by itself (sometimes blurry and unrecognizable); it is the knowledge of who the person is. If you were told that the picture was of someone else, the arousal would fade. Magazines will pay fortunes for a naked picture of an attractive famous person, and nothing at all for a naked picture of someone who looks like that person, even if, on a physical level, it is the very same picture. It's the sexual equivalent of a Vermeer versus a van Meegeren.

Consider also the emerging field of teledildonics, in which one can have sex with a real person over the Web through attachments that provide different types of stimulation. If I were the sort to invest money, I'd invest in that, because I imagine that such an activity, if the technology could be made workable, would be immensely popular. It would provide people with the opportunity to have sex with a real (albeit faraway) person with few of the consequences. It also provides a useful illustration of the "deeper" factors of sexual attractiveness that I discussed earlier. The pleasure one would take in this experience would rest to a large extent on who is pressing the buttons at the other site. A beautiful movie star? Someone of the same sex? Your mom? At a physical level, it might all be the same, but it's not just the physical level that matters.

A final illustration of the essentialist nature of desire comes from a rare disorder called Capgras Syndrome, in which people come to believe that those close to them, including their spouses, have been replaced with exact duplicates. One theory is that it results from damage to the brain areas responsible for the emotional reaction we get when we encounter those we love. A sufferer might then see someone who looks just like his wife, but it just doesn't feel like her. There is the gut feeling that she is a stranger, and so this is resolved by seeing her as somehow an impostor—perhaps a clone, or alien, or robot.

The typical response is fear and rage, and sufferers have sometimes murdered close family members. But there is one exception that I know of, a real-life version of Singer's story of the wandering fool. This is a case study from 1931 of a woman who had complained about her sexual dud of a lover; he was poorly endowed and unskilled. But after suffering brain damage, she met someone "new." He looked exactly like the man she had known, but this one was "rich, virile, handsome, and aristocratic." Sexual and romantic feelings are deep, and her brain damage allowed her to start over, thinking of her lover as a different individual, a better one. This is a vivid example of the essentialist nature of attraction. As Shakespeare put it, "Love looks not with the eyes, but with the mind."

4

IRREPLACEABLE

HOW MUCH MONEY WOULD YOU TAKE FOR ONE OF YOUR kidneys? What about for your baby? How much for sexual intercourse? Suppose a billionaire were arrested or drafted—what would it be worth to you to take that person's place?

People have participated in all of these exchanges for a long time, but they are now illegal in much of the world. In an intriguing discussion titled "What Money Can't Buy," the philosopher Michael Walzer provides a list of blocked exchanges in the United States. These include:

1. People (i.e., slavery)
2. Political power and influence
3. Criminal justice
4. Freedom of speech, press, religion, and assembly
5. Marriage and procreative rights
6. Exemption from military service and jury duty

7. Political offices
8. Desperate exchanges (agreeing to waive minimum-wage laws, health and safety regulations)
9. Prizes and honors
10. Divine grace
11. Love and friendship

These forbidden transactions are "taboo trade-offs." It is not just that we personally don't want to participate in these exchanges or that we believe that if they were permitted people would be worse off in some concrete sense. It's worse. Many people find such exchanges appalling, unnatural, "morally corrosive." In a clever experiment, the psychologist Philip Tetlock and his colleagues presented subjects with stories about a person who deliberates over a taboo trade-off—a hospital administrator who has to choose whether to spend a million dollars to save a dying five-year-old—and they found that subjects disapproved of him regardless of what he ultimately decided. It taints one to think about such choices.

These sorts of trade-offs might seem like exceptional cases. After all, most things do have a price; we have little trouble buying and selling objects such as cars and shirts and televisions. We assign value to such everyday objects on the basis of their utility—what they can do for us. This is what it means to participate in a market economy.

In this chapter, I argue that it's not so simple. I begin by showing just how market *unfriendly* our minds are, how we often reject the notion that objects can be exchanged for money. I then turn to the question of why we like to possess certain things, arguing that while utility is important, there is something more interesting going on. We are essentialists, and so all of us, even young children, think about the things that we own in terms of their hidden natures, including their histories. This essentialism explains what we like about everyday objects—and explains why some of these objects can give us rich and lasting pleasure.

MARKET FAILURES

One summer a few years ago, someone broke into my house, going through the back window we had left open on the ground floor. The window was small, so the thief was most likely not an adult. Next to the window is a desk, and on that desk was a new laptop computer (mine), an older computer (my wife's), and my wallet. The thief took none of these, nor did he (or she, but I'll engage in some profiling here and assume a he) steal the television or DVD player in the room. Rather, he took our Xbox machine and all of our games. Nothing else.

We were baffled by this, and so were the police. The wallet is a particular puzzle, because it was full of money. The simplest explanation, I guess, is that the thief simply didn't notice it. But I can think of a more interesting account.

Perhaps the thief didn't see himself as a thief. The economist Dan Ariely has found that money has a special status. He finds that MIT undergraduates and Harvard MBA students are more likely to steal cans of Coke than dollar bills. This makes intuitive sense. I wouldn't dream of walking into the psychology front office, going into the petty cash drawer, and walking out with $5 so I can pick up something for the children on the way home. I am not a thief. But it feels different to walk into the storage cabinet looking for some other materials, and, by the by, picking up some tape and scissors and paper (net value: $5) to take home for the younger child to do his art project with. I'm not saying that my thief thought he was guiltless, but he might well have inferred that taking cash would have been a whole different level of crime, more hard-core than he wanted to be.

The anthropologist Alan Fiske has developed a framework that helps make sense of this. He notes that there are a limited number of transaction systems across the world. The most natural and universal are Communal Sharing, which occurs within families and some small groups (What's mine is yours; what's yours is mine), and Equity Matching, which involves the exchange of comparable good and services (You scratch my back; I scratch yours). These exchanges even show up in nonhuman primates. The least natural transaction system is Market Pricing. This involves money, debt, interest, higher mathematics, and so on. It might be a wonderfully optimal system, but it is not universal, not shared with other species, and understood only with considerable experience and practice.

These transaction systems trigger different psychologies. Market Pricing—anything having to do with money—is harsh and impersonal, the stuff of law. Ariely's work is one illustration of this. Another comes from my own research—not from the actual findings but from our methods. When a graduate student needs some data from undergraduates, he or she will sometimes sit at a table on campus and ask students to fill out surveys or answer a few questions. Yale students are busy, and often rich, and if we offered them $2, few would stop. Instead we offer Snapple or M&M's. This works better than cash—even though the value of what we offer is less than $2. Money would frame our request as a commercial transaction, and not an appealing one, while the offer of a snack brings out people's better natures.

Similarly, it might be rude to go to someone's house for dinner empty-handed, but it's worse to hand your host a few twenties—or to lean back after the meal, and say, "That was great. Put it on my tab." Money is usually an inappropriate gift, although, using the criterion of efficiency, money is the *perfect* gift. It is better than flowers, wine, or jewelry, because if you give money, the recipient has the option of buying flowers, wine, or jewelry or anything else, or saving it to buy something another day. The problem is that money is for cold-blooded market transactions; for those you like and love, you need to give material things.

There are some exceptions. Money can be a wedding gift, a concession to the financial needs of a newly married couple. (It is less appropriate, though, if the married couple is older or richer than you are.) You can also give money to a child, presumably because the status difference between adult and child is so big that it isn't naturally thought of as an insult.

There are also various workarounds of the money taboo. People can "register" for gifts. Instead of receiving money and buying something with it (taboo), the recipients choose their gifts ahead of time and the givers then buy those items for them (not taboo). In my experience, many married couples do an informal version of this for birthdays and anniversaries: they each tell the other precisely what to buy.

Then there is the gift card, a device that helps the giver (who doesn't have to choose a gift) and the recipient (who gets some choice). The card's similarity to money makes its weirdness screamingly obvious, though: a $50 gift card is just like a $50 bill—except that it can be used in only one store or set of stores, and it will soon expire. From the standpoint of those who sell the gift cards, it is a genius invention—companies earn billions of dollars a year because of unused or expired cards.

We have learned to cope with market exchanges. We are able to put a price on an iPod or a chocolate bar. We do this sort of monetary calculation even for illegal or immoral exchanges; after all, people do sometimes engage in taboo exchanges such as paying for sex, a vote, or a kidney, so they must have some intuition about how much these activities and objects are worth.

More generally, we would be lost if we couldn't assign value to everyday objects and services. This is needed not just for Market Pricing but also for Communal Sharing and Equity Matching. We need to do these calculations when trying to evenly divide up different resources, such as toys, for our children. We don't pay friends for cooking us dinner or taking in our mail, but we do get them gifts, and therefore need to calculate the appropriate value of the gift. Just how expensive should that bottle of wine be? If someone takes care of my dog for a month, it would be insultingly stingy to come back and hand the person a pack of bubble gum (it would be more polite to just give *nothing*), but pathologically generous to buy the person a new car.

Also, in a world of scarce resources, there is a deep sense in which everything has a price. I'm not supposed to put a dollar value on my time with my family, for instance—it is taboo in a Tetlock-like sense to explicitly do so—but apparently I do, because I will leave my family to give a talk to make some money. My wedding ring has sentimental value, and I wouldn't give it to you for $100. But I would hand it over for $10,000.

More grimly, in much of the world, people are forced to make terrible choices, such as women selling themselves for sex to feed their children. Such trade-offs are inevitable even in the richest societies, where governments must balance the value of the environment, housing for the poor, funding for the arts, health care, and so on. Life can be zero-sum, and every penny that goes into supporting an opera company is one penny less for vaccinations for children. Insurance companies calculate how much to reimburse someone for the loss of a toe, an arm, or both eyes. Even people get a dollar value. If the government could save 10 lives at the expense of $10 million (through a vaccination program, say), should it? What about 10 lives for $1 billion? It is impossible to reason about these questions without committing to the most troubling extension of Market Pricing possible—putting a price on human life.

PERSONAL HISTORY

Consider now just those things that we exchange, relatively easily, for money. Not sex and kidneys, but cups and socks. How do we compute the value of such things?

Plainly, there is the utilitarian consideration of what the object can do for you. A car is valuable because it can take you places; a coat can keep you warm; a watch tells time; you can live in a house; a bottle of wine can get you drunk; and so on. These properties are based on the material nature of the objects, nothing more. If someone took my watch and replaced it with a perfect duplicate, its utility as a timepiece wouldn't change.

What's more interesting is that the history of an object also matters. Suppose you ask someone how much she would pay for a coffee cup, and suppose she says $5. You take the money and hand over the cup, and then ask how much she would take to sell the cup back. The rational answer would be $5—or maybe a bit more to pay for the trouble of passing it back and forth. If she sells it for $6, she just made a dollar profit for 10 seconds of work. But the mind doesn't work that way; people usually won't take $6 for it. Its worth increases radically. It's different now. It's *hers*, and this raises its value—a phenomenon known as the endowment effect. Indeed, the longer a person owns an object, the more valuable it becomes.

A different example of the role of personal experience concerns the decisions that one makes about an object. You might think that we choose what we like, which is of course true. But what's less obvious is that we like what we choose.

This was shown over 50 years ago, by the social psychologist Jack Brehm. He asked housewives to rate how much they liked a series of household items, such as coffeemakers and toasters. For each woman, he took items that she ranked as equally attractive, told her that she could take one of them home, and allowed her to choose. After the choice, each woman was asked to rerate the items. Brehm found that the ranking of the chosen item went up and the ranking of the others dropped. (As an aside, ethical standards were different then; when the experiment ended, he told the housewives that he was lying—they couldn't really take home the items. One woman burst into tears.)

You like what you choose; dislike what you don't. There is a simple demonstration of this, the sort one can do in a bar. Take three identical things, such as coasters, and put two of them in front of your subject. Ask him to choose between them. Yes, they are all the same, but still, just pick one. Once he chooses, hand over the chosen object, then bring out the third, and now ask him to choose between the rejected object and the new one. What you'll tend to find is that the rejected object has dropped in value—it is tainted by not having been chosen the first time around, and so the tendency here is to choose the new object.

Nobody really knows why this happens. Perhaps it has to do with self-enhancement; we want to feel good about ourselves, and so we pump up the value of our choices and denigrate the road not taken. Or maybe it is an evolved mental trick to make repeated hard decisions easier—once you choose between two close options, your choice will make the difference between the options seem larger, making it an easier choice in the future. A third proposal is self-perception theory. We assess our own choices as if they were done by another person, and so when I observe myself choosing *A* over *B*, I draw the same conclusion that I would if someone else made this choice—*A* is probably better than *B*.

Whatever the right explanation is, it is clear that one's history with a specific object affects how one values it. This is not limited to adult humans. In a series of experiments in collaboration with the graduate student Louisa Egan (now at the Kellogg School of Management at Northeastern University) and my colleague Laurie Santos, we did a series of choice studies, using the same three-object procedure described above. We found the expected shift in value both with four-year-old children and with capuchin monkeys.

CONTACT

Another relevant aspect of an object is its history before it got to you—where it comes from, what it was initially designed for, who touched it, who owned it, who used it. Sometimes the relevant contact is with someone famous. One can study this in a psychology laboratory, but the phenomenon is obvious when we look in the real world at what people choose to buy and sell.

Just a few minutes on eBay, the online auction site, reveals that contact with a celebrity increases the value of an object. One sort of contact that matters in our culture is a signature. As I write this, Einstein's autograph is $255; an autographed letter by President Kennedy, $3,000; an autographed prison letter by Tupac Shakur, $3,000; a signed poster from the cast of *Star Trek: The Next Generation*, $700. Copies of these signatures are easy to create, impossible to distinguish from originals, and worthless. The originals get their value by dint of their history.

Day-to-day contact with an important person can also add considerable value. In a 1996 auction, for instance, President John F. Kennedy's golf clubs sold for $772,500 and a tape measure from the Kennedy household sold for $48,875. There have been auctions for Barack Obama's half-eaten breakfast (which received a high bid of over $10,000 before it was taken from the site, which does not allow the sale of unpreserved food) and Britney Spears's chewed-up bubble gum. Speaking of Britney, in October 2007, a photographer had his foot run over by her car; he then sold his sock on eBay under "music memorabilia":

> *Authentic sock Britney ran over. The actual sock worn by a TMZ cameraman Thursday when Brit drove over his foot. Tire tread guaranteed authentic!*

This isn't a new phenomenon. In the Middle Ages, there were brisk sales of objects said to be the bones of saints or pieces of the cross upon which Christ was crucified. After Shakespeare died, people cut down the trees around his house to make special lumber for high-priced items. The trees surrounding Napoléon's gravesite were also pulled apart and pieces were brought home as souvenirs. (Napoléon's penis suffered a similar fate, severed by the priest who had administered last rites to him.)

My favorite example of the power of contact is the writer Jonathan Safran Foer's collection of blank paper. Foer began his collection when a friend who was helping to archive Isaac Bashevis Singer's belongings sent him the top sheet of Singer's stack of unused typewriter paper. Foer contacted other authors and asked them to send him the blank pages that they were going to write on next, and he got pages from Richard Powers, Susan Sontag, Paul Auster, David Foster Wallace, Zadie Smith, John Updike, Joyce Carol Oates, and others. He even managed to cajole the director of the Freud Museum in London to hand over the top sheet from a stack of blank paper in Freud's desk. This demonstrates how the most mundane things (blank pieces of paper!) can get value through what one knows about their history.

MAGIC

One theory is that people value these objects because of their intuitions about how they are valued by others. We might pay a lot for a tape measure from the Kennedy household, for instance, because we expect other people to later buy it from us at a higher price or to be impressed that we own it. Another explanation is that these objects are valued because of their power to evoke memories. They remind us of people we enjoy thinking about and are pleasant because of this.

While both of these factors might play a role, neither is sufficient. People often enjoy these objects in their own right, not to boast or make money. Certainly this is true for those personal objects we cherish, such as our children's baby shoes, which nobody else wants and nobody cares that we own. And while it is true that certain objects have positive associations, this does not fully explain the pleasure that they give us. If all I wanted was a reminder of my son as a baby, duplicate shoes would work just as well, and a good video of him would be even better. If someone wants to be reminded of JFK, a giant poster would work fine. There is something else going on here, something that has to do with the contact that these objects have had with special individuals.

Perhaps this something else is magic. The anthropologist James Frazer, in *The Golden Bough*, talks about certain universal beliefs, one of which is Contagious Magic, which "proceeds upon the notion that things which have once been conjoined must remain ever afterwards, even when quite dissevered from each other." Frazer gives voodoo as an example of this: "the magical sympathy which is supposed to exist between a man and any severed portion of his person, as his hair or nails; so that whoever gets possession of human hair or nails may work his will, at any distance, upon the person from whom they were cut."

This sort of theory can explain the appeal of certain objects—through physical contact, they become imbued with an individual's essence. It is not merely that the object brings to mind the idea of the person, then; it is that the object actually retains some aspect of the person.

The most obvious cases are actual body parts. The literary scholar Judith Pascoe notes the pleasure that many collectors get from owning chunks of famous people. Her examples include

Napoléon's penis and intestines, Keats's hair, and Shelley's heart, which was kept close by his wife and ended up being the object of a great custody battle. Pascoe suggests that the Romantic era was a time when people believed that objects, including body parts, were "imbued with a lasting sentiment of their owners." I agree, but I think that this has always been true.

It doesn't have to be an actual piece of a person, though. Something that was once in close contact with the person will do just as well. This explains why money can be made by auctioning off clothes worn by celebrities. It also explains something about the condition of the clothing that people most want. One charity that sells such clothing used to offer a dry cleaning option before sending off the clothes—but they dropped this option because it proved unpopular. People want the clothes as they were when the actors wore them, sweat and all. They don't want the essence to wash off.

In a series of experiments with my Yale colleague George Newman and the psychologist Gil Diesendruck, we tested this positive contagion theory in a more controlled way. We asked our subjects to first think about a living famous person that they admired. (Answers included Barack Obama and George Clooney.) Then we asked how much they would pay for a specific object that was owned and used by this person, such as a sweater. The main focus of this study concerned people's reactions to certain stipulations and transformations. Some of the subjects were told that they were forbidden to resell the sweater or to tell anyone that they owned it. This caused the price to drop slightly, suggesting that one reason they wanted the sweater really does have to do with resale value or boasting rights. Other subjects were told that the sweater was thoroughly sterilized before it got to them. We predicted a much bigger effect here and we got one; there was a drop of almost one-third in how much they were willing to pay. In another study, subjects were told that the celebrity got the item as a gift but never actually wore it—again, this made the sweater less attractive; people would pay less for it. Part of the value of a celebrity-touched object, then, is the implicit notion that it has the residue of the celebrity on it. This finding fits with other research showing that people are more likely to buy a product if it was just touched by someone highly attractive.

We also asked how much pleasure they would get from wearing the sweater. It turned out that having to keep the purchase secret and never selling it had no effect on willingness to wear. But, as predicted, knowing that the object was sterilized or never worn reduces the pleasure that one would get from wearing it.

The discussion so far has focused just on positive contact. But there is the corresponding phenomenon that contact with a reviled person can cause the value of an object to drop. The psychologist Bruce Hood begins his fascinating book *SuperSense* by describing how the city council of Gloucester, England, ordered the destruction of the home of Fred and Rosemary West. This was the house where they raped, tortured, and killed several young girls, burying them under the basement floor and in the garden. The council made a point of removing the bricks, crushing them into dust, and scattering them in a landfill at a secret location. A similar intervention occurred with the apartment where Jeffrey Dahmer lived; it is now a parking lot. In some parts of the United States there are disclosure laws that force realtors to state whether they are selling a "stigmatized home." This effect shows up as well in the laboratory studies of the psychologist Paul Rozin and his colleagues, which find that people are reluctant to try on a sweater worn by Adolf Hitler.

Interestingly, though, there is also a fascination with such negative objects. There are some who would get a kick out of living in an apartment once occupied by Jeffrey Dahmer, putting on Hitler's sweater, or owning a brick from the West household. (This is presumably why the Gloucester council went to the trouble of hiding the remnants of these bricks.) Items such as Charles Manson's hair, paintings by John Wayne Gacy, and the personal effects of Saddam Hussein are routinely sold at "specialty" auctions, sometimes fetching tens of thousands of dollars per item.

This is a minority taste, though. We did a variant of our Clooney/Obama study, this time asking people how much they would pay for a sweater from a *despised* person. Many would pay nothing and said that they would get no pleasure from wearing it. Those who wanted the item didn't care about the sterilization, but if they were told that they couldn't resell it, there was a sharp drop in what they would pay. This suggests that our subjects valued these despised objects in large part because they thought other people would want them.

AN INTEREST IN HISTORY

Do children evaluate objects based on their history? For them to do so, they have to be capable of thinking about objects as distinct individuals. This is not small potatoes. It is far more complicated than responding to object properties. Natural selection can easily wire up a moth's brain to be attracted to the light, or a dog's brain to respond to certain smells, or even a baby's brain to prefer a pretty face to an ugly one. Any simple neural network can generalize, responding in similar ways to similar stimuli. This sort of property sensitivity is so simple that it doesn't even need a brain; even antibodies are category detectors, sensitive to any antigen that has a specific property.

Some scholars have claimed that the brain is nothing more than a generalization machine. We make sense of objects in the world by resonating to the properties that these objects possess. The philosopher George Berkeley nicely summed up this view in 1713: "Take away the sensations of softness, moistness, redness, tartness, and you take away the cherry. Since it is not a being separate from sensations; a *cherry*, I say, is nothing but a congeries of sensible impressions or ideas perceived by various senses."

But Berkeley was wrong. We are not limited to responding to the properties of cherries; we can think about cherries as individual things. You can easily imagine a pair of cherries in a box, each soft, moist, red, and tart, but you know there are two of them, not one. And this is not because we are merely sensitive to the magnitude of the properties—anyone can tell the difference between two small cherries and one big one. You can easily track an individual even if its properties are unstable, as when a caterpillar turns into a butterfly, or a frog into a prince, or when the good people of Metropolis peer at the sky at some vague form and say: "It's a bird, it's a plane . . . it's Superman!" And if one takes a cherry, paints it green, injects it with salt, and freezes it solid, it now has none of the standard properties listed by Berkeley, but it doesn't *disappear*; the individual lives on even though its properties have changed.

Even babies can think about individuals. The psychologist Karen Wynn demonstrated this in an elegant study with six-month-olds. The experimenter shows the baby an empty stage and then blocks the stage with a screen. She then shows the baby a Mickey Mouse doll and places it behind the screen, out of sight. Then she takes another, identical, Mickey Mouse doll and places it behind the screen as well. Then the screen drops. Babies expect two dolls; they look longer, indicating surprise, if one or three appear. This is typically cited as evidence for baby math (they know that 1 + 1 = 2), but it tells us something else as well, which is that babies can track individual objects.

This ability to reason about individuals shows up in children's language by about the first birthday. The initial words of children typically include pronouns like "this" and "that" that can serve to pick out specific individuals in the environment. This is true of children learning every language studied, including Chinese, Danish, Finnish, French, Hebrew, Italian, Japanese, Korean, Quechua, Samoan, and Swedish. Some children create their own pronoun to point out objects around them. At about 12 months of age, my son Max would point and say, with rising intonation, "*Doh?*" He didn't necessarily want us to do anything with the individuals that he was pointing at; he just wanted to show them to us.

THINKING ABOUT individuals is necessary for object essentialism, but it's not enough. Children might be able to tell one thing from another, recognizing that two objects possessing the same properties are nevertheless distinct, but this doesn't mean that they believe that objects have essences or that they think an objects' value can be affected by its history.

To explore this issue, I did a series of studies in collaboration with Bruce Hood. For these studies, we needed a duplicating machine, something that creates perfect copies of real-world objects.

Imagine what one could do with such a machine. One could become rich, copying gold, diamonds, emeralds, and valuable artifacts such as watches and laptop computers. But not all duplicates would be worth the same as the originals. If you copied a stack of bills, you might be tempted to spend the duplicate money, but because history matters in the legal system—a counterfeit is defined as something that has the wrong origin—you could go to prison for a long time. You might put a Picasso into the machine, your wedding ring, or your Tupac signature, but then you would be careful to keep the duplicates separate, since they would be worth much less than the originals. Copying your hamster, dog, or child would have its own special moral and emotional consequences.

We started off small, exploring whether children appreciate, as adults do, that something can be valuable if it once belonged to a famous person. Since our task was a bit complicated, we tested somewhat older children—six-year-olds. Even with this older age group, we immediately ran into a problem—they didn't tend to know any famous people. (Harry Potter didn't count—we wanted someone *real.*) This problem solved itself when Queen Elizabeth II visited Bristol, England, which is where we were running the experiments. We started testing children immediately after the queen's visit.

It was not a serious problem that three-dimensional duplicating machines do not exist. Bruce Hood is an amateur magician, and he found it trivial to create a setup with two boxes in front of a curtain, as shown on the opposite page.

To demonstrate the machine, the boxes were originally open. A green wooden block was placed in one box and both doors were

closed. The experimenter adjusted some controls and then activated a buzzer. Following a delay of several seconds, the buzzer on the second box activated and the experimenter opened both doors to reveal a green block in each box (the "duplicate" block was inserted through the back by a hidden experimenter).

When we showed this machine to children, none thought it was a trick. This fits with other research that finds that children are perfectly credulous about unusual machines. There is no reason why they should be skeptical. They live in a world with giant flying canisters, metal-cutting laser beams, talking computers, and so on. And we already have rudimentary *two*-dimensional duplicating machines—you can take a piece of paper with Michael Jordan's autograph on it, put it in a photocopy machine, press the button, and end up with something indistinguishable from the original. What is so strange about a three-dimensional version of this? For the children we tested: nothing. When asked to explain what they saw, all children said that the machine had copied the block.

We taught the children to give estimates of value, by giving them 10 counters and teaching them to distribute them to pairs of objects based on their value. For instance, they were shown an attractive toy and a rock, and once they agreed that the toy was worth more, they learned to give it more counters.

The children then watched as either a small metal goblet or a small metal spoon was placed into the machine. They were told that this was special because it once belonged to Queen Elizabeth II. After the transformation, the doors were opened to reveal identical objects (goblets or spoons) in each box. Children were then invited to estimate how many counters each item was worth. We had another condition where children were told that the object being duplicated was valuable because it was made of silver; the queen wasn't mentioned.

As we predicted, the queen-owned objects tended to get more counters than the duplicates. Children know that this sort of contact adds value to an object, value that is not carried over to a duplicate. This effect did not occur in the other condition—an object that is special because it is made of silver is no different in value from a duplicate object that is also made of silver. Substances can be duplicated; history cannot.

PEOPLE ARE SPECIAL

People are particularly relevant individuals for essentializing. There is no impulse to think about a rock as interestingly distinct from a similar-looking rock next to it. But it is natural to keep track of individual people. A baby should care very much whether a given woman is really his mother, as opposed to someone who merely resembles her; any mother should have an equally urgent interest as to which baby is hers. And, as discussed in detail in the last chapter, it matters to all of us which specific individual we are sexually or romantically involved with.

Are children especially sensitive to the specialness of social individuals? This question is partially addressed by another set of studies that Bruce Hood and I are currently pursuing with the duplicating machine. In these studies, done with four- and six-year-olds, we duplicate living hamsters. The hamsters are littermates and are thus indistinguishable to the eye. (Well, actually, in one of our studies, one hamster was an enthusiastic eater and bloated up to quite a bit larger than his "duplicate." We replaced him.)

These studies are in progress, but we are finding so far that the children often reject the notion that the duplicate is really a duplicate. That is, while they tend to agree that we have successfully duplicated the physical body of the hamster, they are not always willing to accept that we duplicated the animal's mental states, including what it likes and what it knows. They see the machine as a body duplicator, not necessarily a mind duplicator; the duplicate is a different individual.

Why stop there? What if one built a bigger duplicating machine, with closets instead of boxes, so that a person would walk into one closet and then (by sneaking out through the back curtain) come out the other? What if you did this with the child's mother, setting it up so that it looks like the person coming out of the box is a duplicate, a fake mom? Would the child cringe, draw back in stranger anxiety, scream for the real mother to return?

For ethical and practical reasons we are not quite doing this last experiment. But the writer Adam Gopnik did a milder version using his five-year-old daughter, Olivia, as a subject. When she was out of the house, her fish Bluie died. Gopnik and his wife decided to replace it with a duplicate and brought home a fish indistinguishable from Bluie. But at the last minute, they decided that they did not want to lie to their daughter, at least not entirely, so they made up a compromise story—they told her that Bluie was in the fish hospital for a while, so this was, as a temporary replacement, Bluie's brother. When faced with this identical-looking (and identical-behaving, for that matter) substitute, Olivia was unhappy.

> *"I hate this fish," she said. "I hate him. I want Bluie."*
>
> *We tried to console her, but it was no use.*
>
> *"But, look, he's just like Bluie!" we protested weakly.*
>
> *"He looks like Bluie," she admitted. "He looks like Bluie. But he's not Bluie. He's a stranger. He doesn't know me. He's not my friend, who I could talk to."*

ARMIES IN THE CLOUDS

We have discussed instances in which individuals are special because of contact—typically physical contact—with social beings, such as celebrities and those we love, and instances in which objects are special because they are themselves social beings, animals, or people. In the next chapter, we will deal with a third way in which objects can become special, by being connected in some way with human virtuosity, and this will bring us into the world of art.

Other cases of object valuation have an interesting and unusual status—the individuals are not social beings, but we tend to think of them as if they were. It is old news that humans tend to anthropomorphize, to imbue objects around us with human qualities. David Hume wrote about this in 1757: "We find human faces in the moon, armies in the clouds; and by a natural propensity, if not corrected by experience and reflection, ascribe malice and good-will to every thing, that hurts or pleases us." As one cognitive scientist of religion put it, we have a "hypertrophy of social intelligence."

This can help us make sense of children's attachments to favored objects like teddy bears, blankets, and soft toys, attachments that sometimes live on in adults. The pediatrician and psychoanalyst Donald Winnicott proposed that children use these things as a substitute for their mother (or their mother's breast). He dubbed them "transitional objects" to capture the notion that they are a way station between attachment and independence. This explains a lot. It explains why children get so deeply attached to them and why such objects are soft and cuddly, just like mom. It also explains cross-cultural differences: Japanese children are less prone to have such objects than American children, presumably because they are more likely to sleep next to their moms and so have less need for a substitute.

If such objects are seen as surrogate people, it follows that children should be attached to them as distinct individuals. They should be irreplaceable. Indeed, parents sometimes say that children behave as if this were so, refusing to allow adults to repair their attachment objects and balking at the offer of replacements.

Bruce Hood and I used the duplication machine to explore this issue. We advertised for children with an attachment object.

To count as such an object, several requirements had to be met, including that the child had to sleep regularly with the object and had to have possessed it for at least one-third of his or her life. Parents brought their children into the lab along with their object. As a comparison group we also brought in another group of children who didn't have attachment objects; these parents were just asked to bring in any particular object that the child liked, such as a favorite small toy.

The children were between three and six years old. The study was simple. Once they were in the lab, we showed them the duplicating machine and demonstrated it. Then the experimenter suggested that they copy the child's own object. If the child agreed, the experimenter placed the object in the box, duplicated it, and (with the two boxes closed) asked the child which object he or she would like to keep.

When the children without attached objects put their toys in the box, most chose the duplicate. It was cool because it was a machine-created copy. They were disappointed when we explained to them that it was all a trick and that their object wasn't really duplicated.

The children with attachment objects behaved differently. Some of them refused to let the experimenter put their object into the duplicating machine at all. Of those who allowed it, most preferred to take home the original.

When this work was discussed in the popular press, Bruce got the following letter:

Dear Dr. Hood— . . .

My 86 year old mother still sleeps every night with the little pillow from her baby cot. She has been apart from it for one night only in 86 years and that was when she forgot to take it down into the bomb shelter during an air-raid. She has stipulated that it be buried with her. It even has a name, Billy.

I don't think she would swap it for a copy.

Most objects are not like Billy. We are willing to part with them or replace them with copies. But everything is either a social being or has been in contact with a social being, and so even the most mundane things have histories. This is their essence. And for some of these objects—like Billy or Bluie, Kennedy's tape measure, George Clooney's sweater, Napoléon's penis, or the shoes that my child wore as a baby—this essence is a source of great pleasure.

5

PERFORMANCE

IN THE MORNING OF JANUARY 12, 2007, A YOUNG MAN IN jeans, a long-sleeved T-shirt, and baseball cap walked into a Washington subway station and pulled out a violin. He laid out his violin case in front of him, seeded it with a few dollars and some change, and then played six classical pieces for the next 43 minutes, as over a thousand people walked by.

This was no ordinary street performer. He was Joshua Bell, one of the world's great violinists, and he was playing his $3.5 million violin, handcrafted in 1713 by Antonio Stradivari. A few nights before, Bell performed at Boston's Symphony Hall. Now he stood in front of commuters, playing for coins. This was an experiment by Gene Weingarten, a reporter for the *Washington Post*. It was intended as an "unblinking assessment of public taste": how would people respond to great art in a mundane context, when nobody was telling them how great it was?

The people failed. Over a thousand commuters passed, and Bell netted a bit over $32. Not bad, but nothing special. The commuters were indifferent to what they were hearing. Weingarten spoke to Mark Leithauser, a senior curator at the National Gallery, who put this indifference in a broader context:

> *Let's say I took one of our more abstract masterpieces, say an Ellsworth Kelly, and removed it from its frame, marched it down the 52 steps that people walk up to get to the National Gallery, past the giant columns, and brought it into a restaurant. It's a $5 million painting. And it's one of those restaurants where there are pieces of original art for sale, by some industrious kids from the Corcoran School, and I hang that Kelly on the wall with a price tag of $150. No one is going to notice it. An art curator might look up and say: "Hey, that looks a little like an Ellsworth Kelly. Please pass the salt."*

As Weingarten puts it, Joshua Bell in the subway was art without a frame.

At the very end of the performance, Stacy Furukawa passed by. She had been at one of Bell's concerts a few weeks before, and stopped 10 feet away from the musician, grinning and confused. When he was finished, she introduced herself and handed over $20. Weingarten did not count this as part of the total—"it was tainted by recognition." Furukawa's gift was because of the man, not (or not entirely) because of the music.

This experiment provides a dramatic illustration of how context matters when people appreciate a performance. Music is one thing in a concert hall with Joshua Bell, quite another in a subway station from some scruffy dude in a baseball cap.

It is a clever demonstration, but perhaps not surprising. Everyone knows that the value of a painting shoots up if it is discovered to be by a famous artist, and plummets if it is discovered to be a fake. *The Night Watch* is the most famous painting in the Rijksmuseum, but if it were discovered tomorrow to be a forgery, its value would go from priceless to bupkis. Origins matter.

This might seem irrational. If you liked *The Night Watch* when you thought it was by Rembrandt, why should you like it any less if it turns out to be by Joe Shmoe? If you would pay dearly to listen to the performance of Joshua Bell, you should enjoy the same performance by a stranger. It is the same paint on canvas; the same sequence of sounds. To respond otherwise reveals human weakness, some blend of snobbery, groupthink, and intellectual laziness.

This was the view of Arthur Koestler, who, in his 1964 book on creativity, *The Act of Creation*, tells a story about a friend he calls Catherine. She received as a gift a drawing that she took to be a reproduction of a Picasso, from his classical period. She liked it, and hung in on her stairwell. But when Catherine had it appraised and it turned out to be by Picasso himself, she was delighted, and moved it to a more prominent part of the house. Catherine insisted to Koestler that she now sees the artwork differently. It looks better to her.

Koestler is annoyed: "It proved quite useless repeating to her that the origin and rarity-value of the object did not alter its qualities—and accordingly, should not have altered her appreciation of it, if it has really been based on purely aesthetic criteria as she believed it to be." He goes on to say that it would be fine if she just admitted that she just gets a kick out of owning a Picasso. What really bothers him is that she insists that the artwork is now more beautiful than when she thought it was a reproduction.

For Koestler, Catherine is a snob. A snob is someone who applies an inappropriate standard. A social snob is someone whose choice of friends is guided by their status, not their deeper qualities. Koestler tells us about a sexual snob, a young woman from Berlin, in the days before Hitler, who would have sex with any author, male or female, so long as his or her books had sold more than 20,000 copies. Koestler finds this ridiculous: "the *Kama Sutra* and the best-seller list were hopelessly mixed up in her mind." For him, Catherine is an art snob. She gets pleasure not from the artwork itself, but from knowing who created it.

Before Koestler, the Dutch forger Han van Meegeren would have agreed. He hated modern art and started his career producing paintings in the Rembrandt style. He was unsuccessful and had terrible luck with critics, one who said, with excellent prescience, that "he has every virtue except originality."

Partially as an act of revenge—and partially to get rich—he started to paint Vermeers. The critics raved. *The Supper at Emmaus* was perhaps the most famous painting in Holland. The leading critic of Dutch baroque art swooned: "We have here a—I am inclined to say *the*—masterpiece of Jan Vermeer of Delft." Van Meegeren, who was quite the egomaniac, would visit this painting in the Boijmans Gallery and loudly tell other visitors to the museum that it was a fake, just to hear them tell him that this was nonsense, only a genius like Vermeer could paint so well.

He might never have been caught, but he was arrested for selling a Vermeer to the Nazi Hermann Goering and charged with treason. He then confessed that it wasn't a Vermeer that he had sold, it was a van Meegeren—and many other Vermeers were van Meegerens as well.

I began this book by describing this episode from Goering's perspective, but think now how humiliating it must have been for the critics. Admittedly, some had their doubts at the time, and some contemporary critics find it hard to believe that anyone could have ever been fooled. (Among other complaints, one of the faces in *The Supper at Emmaus* looks suspiciously like the actress Greta Garbo.) But many critics at the time rhapsodized about the beauty of these paintings. They recanted, though, once they discovered who the artist was. As one expert wrote, "After Van Meegeren's exposure, it became apparent that his forgeries were grotesquely ugly and unpleasant paintings, altogether dissimilar to Vermeer's."

We may be living through a similar case right now. A couple of years ago, Sotheby's sold *Young Woman Seated at a Virginal* for $32 million after a long debate over who painted it. The experts decided it was Vermeer, hence the price, but if they turn out to be wrong, as some think, its value will plummet, and there will be, again, some very embarrassed art experts. Presumably some of them will conclude that the painting is not as lovely as they thought it was.

If it turns out to be a forgery, *Young Woman Seated at a Virginal* might end up in the Bruce Museum, in Greenwich, Connecticut, just an hour drive from my house. As I write this, this is where you can find *The Supper at Emmaus*, as part of a special exhibit on fakes and forgeries. Thus is a small and pleasant museum, and it occurred to me, as I stood in front of the painting, that I could pull it from the wall, march past the elderly woman at the entrance, place it carefully in the back of my minivan, and take it home. If I had pulled this off in early 1945, I would have committed one of the great art thefts of all time. Now this would be a joke; the headline would be: "Deranged professor steals worthless painting."

What has changed? Why does this forgery give us so much less pleasure? This chapter will try to answer that question. It starts with paintings and music, moves to art more generally, and then turns to related pleasures such as sport. I am going to suggest that our obsession with history and context—what we see in the Bell experiment, in the Catherine story, and the rise and fall of *The Supper at Emmaus*—is not snobbery or silliness. Much of the pleasure that we get from art is rooted in an appreciation of the human history underlying its creation. This is its essence.

EAR CANDY

As with the other pleasures that we have discussed so far—sex, food, and consumer products—I admit that some of our response to music and painting is not, in the sense I'm interested in here, deep. Some things are simply good to listen to or look at for reasons that have nothing to do with essentialism or history or context.

This does not mean that we know what these reasons are. In 1896, Darwin described the love of singing or music as one of the most mysterious features of humans. It still is. It is no mystery why we enjoy food, water, sex, warmth, rest, safety, friendship, and love—these are good things to have, survival-wise and reproduction-wise. But why would we so enjoy certain rhythmic series of sounds? Why do humans, everywhere, devote so much of their time and energy to song and dance? Among the Mekranoti people of the Amazon, the women sing one to two hours per day and the men sing two hours or more at night. They live a subsistence lifestyle, yet they spend hours singing! This looks like a perfect waste, so superfluous that it might tempt you away from evolutionary biology and toward a belief in divine intervention. Kurt Vonnegut gave this as his epitaph: "The only proof he needed for the existence of God was music."

Music is a uniquely human pleasure. It might soothe the savage breast, but only the human breast, not that of a rat or a dog or a chimp. Perhaps there are counterexamples to this—if you tell me that your cat is transfixed by your guitar playing, who am I to argue?—but there is no experimental evidence that any nonhumans show a preference for musical sounds. One way to test this is by putting animals in a maze in which different locations correspond to different sounds; one can determine what the animals like by watching which location they go to. Using this method, researchers find that primates such as tamarins and marmosets prefer silence to lullabies, and they show no preference for consonant music versus dissonant music. Monkeys don't care about whether you are exposing them to rock music or to the sound of fingernails screeching against a chalkboard.

In contrast, just about all humans like music. It is harder to test babies than monkeys, because you can't make babies run down arms of a maze, but other methods can determine their preferences. One procedure is to have babies turn their heads to listen to sounds, monitoring what sounds they prefer to attend to. Tested this way, babies prefer consonant music over dissonant sounds, and they enjoy the sound of lullabies. This pleasure in music continues through life; while the extent of this pleasure varies hugely, only the brain-damaged are indifferent.

• • •

THE PSYCHOLOGIST Steven Pinker describes music as the human universal that shows the clearest sign of being an accident. For him, music is "auditory cheesecake," an invention that tickles the brain like cheesecake tickles the palate: "Cheesecake packs a sensual wallop unlike anything in the natural world because it is a brew of megadoses of agreeable stimuli which we concocted for the purpose of pressing our pleasure buttons." Pinker argues that this is true for the arts in general, with the possible exception of fiction.

What evolved pleasure buttons might music press? Pinker discusses several possibilities, including language, which shares with music the unusual property of being rule-based and recursive, with the power to take a limited stock of units (for language, words or morphemes; for music, notes) and combining them into a potentially infinite number of hierarchically structured sequences. But there are also differences. Language is a system for expressing meaningful propositions, which is what is happening right now as you read this. Music can convey emotion (think of the feeling of tension evoked by the theme song of *Jaws*) but it's a dud as a communication system, unable to convey the simplest of propositions. Music gives pleasure through its sound, language usually doesn't—we typically enjoy language because of what is said, not how it is said. On the other hand, there is a pleasure to singing, which combines music and language, and every baby enjoys the sound of its mother's voice.

Other scholars propose that music is an adaptation. This does not deny that the pleasure of music is to some extent built up from other, evolutionarily prior, parts of the brain. In biology, everything new comes from something old. But to say that music is an adaptation is to make the further claim that it exists because it conferred to our ancestors a reproductive advantage. The genes of those who created and enjoyed music out-produced those who didn't.

The most prominent modern defender of this view is the psychologist Daniel Levitin. He suggests that synchronous song and dance evolved as social adaptations. Music can help coordinate war parties, it can make collective tasks easier to do, and, most of all, it can establish emotional bonds with other people. If he is right, then the story of the evolution of music would be much the same as the story of the evolution of other traits that connect people with their groups, such as feelings of solidarity and community.

Even if music has these social advantages, though, making the adaptationist case would require evidence that it exists because of them. The adaptationist has to make plausible the idea that our ancestors without music were less accepted by their fellows and less lucky with mates than their musically savvy neighbors. Further, one needs to explain how the specific features of music evolved. If synchrony is what matters, for instance, why don't we just grunt, shout, and scream in unison? Why are we so moved by the complexities of music, by tones, chords, and so on?

In any case, Levitin's account captures something deep about musical pleasure, something that has been missed by many scholars. This is the importance of movement. Most languages have a single word for both singing and dancing, and when people listen to music while perfectly still, parts of the motor cortex and cerebellum—the segments of the brain that have to do with moving around—are active. This is why we so often rock to music, an impulse that can be irresistible to a child. It would be a scientific misstep, then, to develop a theory of music that took the solitary and still appreciation of tonal patterns as the central phenomenon that has to be explained. This would be like a theory of sex that only studied phone sex or a theory of food preference built around research on people with no sense of smell.

Indeed, there is evidence that if you move in synchrony with other people, you like them more, you feel more connected to them, and you are more generous to them. Song and dance is the ultimate team-building exercise. Most of us are familiar with the emotional rush of linking arms and dancing at a Jewish wedding, or being at a rave, or in a pub with drunken friends. One can experience this secondhand by watching others sing and dance, as in the famous YouTube video "Where the Hell Is Matt?," in which a cheerful nondescript American dances with people around the world. This effect of music may explain why religion has so much singing and chanting and dancing: it establishes solidarity with those of your faith.

Claiming that song and dance are adaptive because they connect you to your community just pushes the question back—why are we so constituted to feel close to those we sing and dance with? Nobody really knows. There are adaptationist accounts, but I wonder if it's due to a glitch in the system. If I dance with others, and they move with me, their bodies moving as I intend my own body to move, it confuses me into expanding the boundaries of myself to include them.

WE HAVE looked so far at music in a general way. Hopelessly general, perhaps. The music one likes is determined in part by the music one is surrounded by—the Top 40 in India are not the same as those in the United States. Even within a country, individuals vary. In my own family, there are sharply contrasting views about bluegrass, thrasher music, classic rock, and opera, and long car rides require careful negotiations over who gets control over the radio.

We know that some of our tastes are established early in development. In one experiment, mothers played certain musical pieces (pieces from Vivaldi, songs by the Backstreet Boys, and so on) to their babies in the womb and didn't play those pieces again until the babies' first birthday. This experience had an effect—the one-year-olds tended to prefer the music that they had heard before they were born.

People spend much of their waking lives passively listening to music. You might expect, as a very simple hypothesis, that we tend to enjoy the music we hear the most—a version of the "mere exposure" effect we talked about earlier regarding sex. Familiar is good. The problem, though, is that too familiar becomes cloying and unpleasant. A rule of pleasure is that it is an inverted U (a ∩)—when you first experience something, it's hard to process and not enjoyable; upon repeated exposure, it's easy to process and gives pleasure; then it gets too easy, and therefore boring or even annoying. We might be cautious about a food at first, then eat it frequently and with joy, but few people would enjoy eating the same main course for a thousand meals in a row. For music, the middle peak of the inverted U can last a while, but any song will become unbearable if you hear it often enough. The shape of the curve is stretched and squeezed by the complexity of the music. A complex piece might take a long while to like and then a long time to get sick of; usually something like "Mary Had a Little Lamb" will go through the curve a lot quicker.

• • • •

ANOTHER FACTOR in determining how much you like a song, or a musical genre more generally, is how old you are when you first hear it. In 1988, the neuroscientist Robert Sapolsky did an informal experiment to look at this, contacting radio stations and asking them when most of the music that they play was first introduced and what the average age of their listeners was. He found that most people are 20 or younger when they hear the music they're going to want to listen to for the rest of their lives. If you are older than 25 when some new form of music is introduced, you are unlikely to enjoy it. As Sapolsky put it, "Not a whole lot of seventeen-year-olds are tuning in to the Andrew Sisters, not a lot of Rage Against the Machine is being played in retirement communities, and the biggest fans of sixty non-stop minutes of James Taylor are starting to wear relaxed jeans."

Why? There is a temptation to go for a simple neurological explanation here. Our brains start off all loose and flexible, and then harden up. But as Sapolsky points out, it's not as if there is a general loss of openness to new experience at this period; the window for new musical taste is open at a different period of life from those for other tastes such as food.

Levitin has a better idea. Music is social, and the locking in of musical preference is linked to the time of life where you affiliate yourself with a certain social group—where you come to a decision as to what sort of person you are. This happens late in modern Western societies: roughly in one's late teens and early twenties, which matches what Sapolsky finds. This sort of social theory might explain something else, which is that out of all the music that young people are exposed to, they prefer the styles that are most recent. This is because they want to be sure to affiliate with their contemporaries, and, as the economist Tyler Cowen puts it, "The problem with old music is simple. *Somebody else already liked it.* Even worse, that somebody else might have been one's parents."

EASY ON THE EYES

Just as with music, some visual pleasures are superficial. When it comes to visual art, some colors and patterns just catch the eye in a certain way. Parents aren't wasting their time when they paint bright colors and systematic patterns on the walls of their babies' rooms. Babies like this sort of thing, and this has nothing to do with essentialism.

Such preferences are studied as part of a field known as experimental aesthetics. Psychologists create different shapes, such as polygons, manipulate them on parameters such as typicality and symmetry, and ask subjects what they prefer. Such studies support the ∩-shaped framework discussed above, as well as the easy-on-the-eyes explanation of pretty faces discussed in the sex chapter—people like images that are easy to process.

In the end, though, this research provides a fragile foundation for the study of visual pleasure. It is hard to see how you can learn about visual pleasure by showing people visual patterns that they wouldn't otherwise choose to look at. This is not to disparage the research—it might be interesting for other purposes to learn why people prefer one geometrical figure to another. But we don't pay money and give up time to look at black and white polygons on a computer screen. These don't give us pleasure.

What does? One immediate observation is that people like to look at realistic pictures of peaceful domestic scenes, flowers and food, attractive landscapes, and, most of all, people—including those we love and those we admire. If you are in your home or office now, my bet is if you look at the human-made images around you, you'd find exactly those sorts of representations. Your screensaver might be a forest or a beach, there may be photographs of loved ones on your desk, and so on. There is plenty of art that doesn't fit this mold, and I'll turn to the Pollocks and such later on, but it is worth noting just how many of the representations around us are attempts to mimic the sorts of things that we enjoy looking at in real life.

A simple illustration of this is pornography. Many people enjoy looking at attractive naked people, for mundane Darwinian reasons. But there aren't always attractive naked people around when you need them. So we have created two-dimensional surrogates that simulate the experience, and hence inspire much the same reaction of lust that would be inspired by the real thing. You are not responding to the surrogate as an artwork, you are responding to the naked woman (or man, or couple, or threesome, etc.) that is represented.

The pleasure of sham nudity isn't limited to humans. In a recent study, male rhesus monkeys were put into an experimental setup where they could choose, by moving their heads, to either receive some sweet fruit juice or to get to look at a picture. There were two sorts of pictures that monkeys would give up the juice for—female hindquarters and the faces of high-status male monkeys. Two major vices—pornography and celebrity worship—are not exclusively human.

IT USED to be thought that an appreciation of realistic pictures requires learning, and there are tales of anthropologists discovering primitive tribesmen who couldn't make any sense of pictures because they had never seen any before.

This was proven false by a clever experiment reported in 1962. The psychologists Julian Hochberg and Virginia Brooks took a child and raised him without any access to pictures until he was 19 months old. (They didn't say so in the article, but the child was their own.) Then they showed him photographs and line drawings of familiar objects and asked him to name them. He did so easily. More recent studies find that even babies have some tacit grasp of the correspondence between a realistic picture and the object it depicts. If you let a five-month-old play with a doll and then take it away and show the baby two pictures, one of that doll and the other of a different doll, she will look longer at the picture of the unfamiliar doll, showing some sensitivity to the correspondence between the familiar doll and its picture.

Children can be so gripped by pictures that they treat them as the things themselves. Observant parents have noticed weird behaviors such as their children trying to step into a picture of a shoe or scratching at pictures to try to get at the depicted object, and careful experimental studies have found this behavior not just in the United States, where children have plenty of experience with pictures, but also among illiterate and poor families in the Ivory Coast, where pictures are rare. Children can tell pictures from real things (they are far less likely to reach for pictures), but the lure of resemblance is sometimes hard to resist.

This is true for adults too. Sometimes we find it hard not to think of a representation as the very thing that it represents. The psychologist Paul Rozin and his colleagues did a series of lovely studies in which they asked people to hold rubber vomit in their mouths or eat fudge that is baked in the shape of dog feces. This is hard for many of us—it *is* vomit; it *is* feces. My colleagues and I recently did a series of studies in which we took pictures of people's precious objects—their wedding rings, say—and asked them to cut the pictures up. They were willing to do so, but measures of skin conductance showed that they were in a state of mild anxiety, as if they were destroying the precious things themselves. And if you ask people to throw darts at pictures of babies, they tend to miss.

DISPLAYS AND FITNESS

Here are some reasons one might value a painting:

1. It could be attractive in a low-level way; the patterns might please the eye.
2. It could resemble something attractive, like colorful flowers or a beautiful face.
3. It might be familiar. This is the mere exposure effect: up to a point, familiarity breeds pleasure.

    Does mere exposure really work for paintings? The psychologist James Cutting asked why some French impressionists are preferred to others. In one study, he found that adults, but not children, typically prefer paintings that have been frequently published in the last century over those that are more obscure.

    Of course, this might not be an effect of familiarity. Maybe it's the other way around: Perhaps the paintings that are frequently published are *better* than the rest, and the adults are responding to this superior quality, not the frequency. Cutting tested this in a second study in which he presented the more obscure impressionists to his undergraduate class on visual perception, showing them for a few seconds each, without any comment, as part of PowerPoint presentations on other topics. Then he tested his students at the end of the semester, asking which paintings they liked the most. It turns out that this brief exposure flips the normal preference pattern. Now they prefer these obscure paintings to more famous ones, just because of the brief exposures. Simply seeing a painting, then, makes you like it more.

4. It might be associated with a positive memory. This matters a lot for photographs, which display experiences such as weddings, graduations, and reaching the top of Mount Everest.
5. It might nicely complement a room. (After all, the shape of a painting does influence its price.)
6. It might enhance your status, impressing those who see that you own it. A painting of Christ on the cross or of the Sabbath table advertises your piety. A modernist work shows how much you know about and care about art. A provocative piece can signal religious or sexual sophistication. And an original by a famous artist is an easy and nonblatant way of telling everyone how rich and successful you are. There is certainly pleasure in that.
7. It might gain value from positive contagion. As discussed in the last chapter, the mere fact that an object has touched someone famous and esteemed can enhance its value.

This list misses a critical factor, though. We are also interested in how paintings are created, and we get pleasure from what we infer to be the nature of that creative process.

This is the view that the philosopher Denis Dutton defends in his important book, *The Art Instinct*. For Dutton, part of the appeal of works of art is that they serve as "Darwinian fitness tests." He develops this claim in the context of a sexual selection theory of the origin of art, one proposed originally by Darwin and then extended and developed by the psychologist Geoffrey Miller, whose work we discussed earlier in the context of sex. According to this theory, art is the peacock's tail. It has evolved as a signal of Darwinian fitness, to attract mates.

As the best animal analogy to human art, Miller turns to the bowerbird, a species found in New Guinea and Australia. Males are the artists. They fly around and collect colorful objects such as berries, shells, and flowers, and bring them back and arrange them in symmetrical and complicated patterns—bowers. Females are the discerning and brutal critics. They check out the bowers, looking for the most creative one, and they mate with its creator. A successful male might mate with 10 females; an unsuccessful one will be celibate. After mating, the female flies off to lay the eggs and the male never sees her again. The life of a successful male bowerbird is a lot like that of Pablo Picasso.

This all suggests a sexual selection account of bowerbird creativity—females are sensitive to bower building as an indicator of fitness, such as intelligence, skill, discipline, and so on, traits that are useful to have in offspring. Miller and Dutton propose that the human artistic urge has been shaped in the same way. Good art is difficult to do. A good artist is a good learner and planner, exhibits intelligence and creativity, and is successful enough at surmounting the immediate obstacles of life (food, shelter, and so on) to find the time and resources for these nonutilitarian inventions. Females like this sort of thing: female bowerbirds are drawn to it when they see it in male bowerbirds and female humans are attracted to it in male humans. The motto here comes from Renoir: I paint with my penis.

Darwin has a similar view in his discussion of music, suggesting that it emerged in part "for the sake of charming the opposite sex." It is not hard to see how song and dance can impress—developing and maintaining a rhythm for an extended period is a useful display of intelligence, creativity, stamina, and motor control, all of which are positive traits in a mate. And one does not need to do an empirical study to show that, in modern times, highly successful musicians, from Mick Jagger, to, well, Joshua Bell, have little problem finding mates. Musical skill is attractive, and a sexual selection theory is a plausible candidate for explaining its origin.

But while I agree with Dutton and Miller that art can be a fitness display, and that both the urge to create and the pleasure we get from such creations may have been shaped to some extent by sexual selection, there are problems with this theory as an explanation for why we like art.

For one thing, the obvious fact about the peacock's tail is that peacocks have them, not peahens. It is the male bowerbirds who create bowers and the females who rate them. This is how sexual selection works. Because the cost of sex is usually greater for the female than for the male (for most animals, females raise the offspring, while men contribute nothing but sperm and a bit of time), the selection goes only in one way—men compete for female attention, females judge the males.

This is a poor model for people. Miller argues that men are more motivated to create art than women and that women are more attuned to appreciate it than men are. This might be true to some extent, but, still, in societies in which everyone is given the chance to create art, there is no shortage of female poets, writers, painters, singers, and the like.

A fair response to this is that people aren't peacocks; we are a relatively monogamous species, and for us sexual selection might go both ways, with everybody in the business of showing off fitness traits and evaluating the traits of potential mates. Still, the problem remains that the appreciation of art often has nothing to do with sexual interest. A man doesn't have to be gay to admire Picasso. Children, who are not yet involved in the business of attracting mates, are among the most enthusiastic artists of all, and the elderly, including women long past their reproductive years, get pleasure from creating and admiring art.

There is also a more general concern. A sexual selection account can explain why Picasso was so successful in finding mates—his creations were testaments to his Darwinian virtues. But this explains the attractiveness of the artist, not the pleasure that people take from the *art*. Why do people enjoy Picasso's paintings so much, though the man himself is long gone?

CONSIDER NOW a modified theory, with two parts.

First, displays of cleverness, discipline, strength, speed, and so on capture our interest because they reveal relevant properties of an individual.

Isn't this the Miller and Dutton argument? Yes, in part. Men do generate displays in the hopes of getting women to mate with them, and women do assess male displays to find a mate with top-of-the-line DNA. But it is not just sex. We also evaluate individuals for their qualities as friends, allies, and leaders. Indeed, although this is cold-blooded, we often have to evaluate the qualities of our children to see who has the best chance of survival and future reproduction. In the novel *Sophie's Choice*, William Styron's character is forced to choose whether her younger daughter Eva or her blond, blue-eyed son Jan would be gassed in Auschwitz, and she chooses to sacrifice Eva, a savage but logical decision given that Jan has a better chance of surviving the camp. Even in a world of plenty, milder versions of such dilemmas remain: parents are often in the position of allocating resources to their children and don't always choose perfectly equitable divisions. It is in the children's best interest, then, to impress Mom and Dad with their sterling qualities.

This is a good place to emphasize that fitness evaluation is a claim about how art evolved, why we are compelled to create it and enjoy it. It is not a claim about psychological motivation, either conscious or unconscious. When a child proudly shows off her drawing to her father, she is not saying to herself, consciously or unconsciously, "This will impress him, and he will give me more food than my brother." When you admire a painting, you don't usually think, "The virtuosity and skill of that work indicates an artist with excellent qualities; I will attempt to mate with or befriend him." Evolutionary functions have nothing to do with psychological motivation. William James pointed this out long ago using the example of food, when he noted that not one man in a billion thinks about utility when eating: "He eats because the food tastes good and makes him want more. If you ask him why he should want to eat more of what tastes like that, instead of revering you as a philosopher, he will probably laugh at you for a fool."

The second part of the theory is that we have evolved to take pleasure from virtuoso displays. This motivates us to seek out such displays, it drives us to create them ourselves, and it provides a psychological mechanism underlying our attraction to artists—they are capable of creating these objects that give us so much pleasure, and we tend to like those who give us joy.

If paintings and other static artwork are displays, then they are understood and appreciated in part based on how we think they were created. The historical nature of artwork has been emphasized by many scholars but was argued most forcefully by Dutton over 25 years ago:

> *As performances, works of art represent the ways in which artists solve problems, overcome obstacles, make do with available materials. The ultimate product is designed for our contemplation, as an object of particular interest in its own right, perhaps in isolation from other art objects or from the activity of the artist. But this isolation which frequently characterizes our mode of attention to aesthetic objects ought not to blind us to a fact we may take for granted: that the work of art has a human origin, and must be understood as such.*

The idea, then, is that certain displays—including artwork—provide us with valuable and positive information about another person. We have evolved to get pleasure from such displays. This is another instance of essentialism at work, another case in which objects are thought to have invisible essences that make them what they are. For something such as a hunk of meat the essence is material; for a human artifact such as a painting, the essence is the inferred performance underlying its creation.

Do people really think that works of art have invisible essences that are rooted in their history? I think that even young children do. I first became interested in the psychology of art over a decade ago, when my two-year-old son smeared paint on cardboard, and then proudly told me that it was "an airplane." This was surprising to me as a developmental psychologist, because the consensus in my field was that children's naming is based on appearance: for a child, the word "airplane" should refer to something that looks like an airplane. But Max's picture didn't look like one at all; it was a colorful blob. He wasn't unique in his behavior. A quick check of the literature revealed that it is entirely typical for children to create pictures and name them—a doggie, a birthday party, Mommy—even though their pictures resemble none of these things.

Along with my graduate student Lori Markson (now at Washington University in St. Louis), I explored the idea that these names aren't driven by what the pictures look like; they are chosen on the basis of the pictures' histories. It was an airplane because Max had wanted it to be an airplane. This was supported in a series of studies that found that even three-year-olds would name their pictures based on what they were intending when they created it. We also found that the same holds for pictures that other people draw. If a three-year-old watches someone stare at a fork and draw a scribble, she will later name the scribble "a fork"; if the same scribble was created while the person looked at a spoon, she will call it "a spoon." In more recent studies with a postdoctoral fellow, Melissa Allen (now at Lancaster University), I found that even 24-month-olds are sensitive to a drawing's history when deciding what to call it.

PERFORMERS

This focus on history helps explain why we prefer originals.

In the last century, philosophers have argued that the emerging technology of reproduction will make this preference go away. Walter Benjamin suggested that "for the first time in world history, mechanical reproduction emancipates the world of art from its parasitic dependence on the real." André Malraux argued that originals would no longer be important, because every museum could contain all the artwork in the world. Why do we even need museums? Think about Bill Gates's home in Seattle, with large screens on the walls that display visual artwork. Then imagine such screens in every house, capable of reproducing any painting you would ever want to see.

But there is, by definition, only one original, so there will always be a certain status associated with seeing it or, much better, owning it. Also, the original has been in contact with the artist, and, as discussed in the previous chapter, this sort of positive contagion is attractive. Most of all, the original has a special history, as it came into existence through a creative process, far more impressive than the technical skill of a forger. It is our sensitivity to this history that explains why the love of originals isn't going to go away.

The focus on performance can also help us make sense of disagreements about art. Take the nonrepresentational paintings of Jackson Pollock. Many people are unimpressed with these. This negative reaction is in part because they are not obvious skill displays. They look easy; the classic line here is: "My child could do that." The art educator Philip Yenawine objects to this position. He responds by describing one of Pollock's works—*One (Number 31, 1950)*. He notes its great size (about 9 feet high, 17 feet wide) and marvels at the technical and imaginative problems involved in creating sweeping arcs of paint, maintaining the separate elements, and so on. If you think this is easy, he remarks, why don't you give it a shot yourself?

This disagreement over Pollock, then, is in part a disagreement about history. If Yenawine could convince the skeptic just how hard it is to make these paintings, the skeptic might grow to have a better appreciation of the work. Similarly, if Yenawine saw a six-year-old slapping paint onto a giant canvas for 10 minutes, creating something that was indistinguishable from *One (Number 31, 1950)*, I bet he would never again get so much pleasure from Pollock's art.

What do we look for when we assess performance? One crude but relevant consideration is the perceived amount of effort. The psychologist Justin Kruger and his colleagues tested this in a straightforward way, by exposing subjects to a poem, a painting, or a suit of armor and telling them different stories about how long they took to make. For instance, subjects would be shown an abstract artwork by the painter Deborah Kleven; half would be told that it took 4 hours to paint, and half would be told it took 26 hours. As predicted, those who were told that it took more time to create provided higher ratings for quality, value, and liking.

(I suspect that this explains a tidbit about art pricing: size matters. For a given artist, the bigger the painting, the more it tends to cost. This might reflect the intuition that it's harder to paint a large painting than a small one. More effort leads to greater pleasure leads to greater value.)

Effort matters as well for how much we value our own creations. When instant cake mixes were introduced in the 1950s, they were unpopular—until the manufacturers changed the recipe so that the housewives had to do some work on their own; they had to add an egg. This made it a better product. This enhancement of value through one's own effort is what the psychologist Michael Norton and his colleagues call "the IKEA effect," after the popular Swedish furniture store where you usually have to assemble the products yourself. They demonstrate this in the lab, finding that subjects value their own creations, simple origami frogs, more than the same objects created by others.

Effort is one factor, but not the most important one. After all, when we prefer the Vermeer to the van Meegeren, it's surely not because we think that Vermeer worked longer and harder. What matters more is our intuitions about creativity and genius.

A striking example of how these intuitions matter was when Marla Olmstead became famous for her abstract artwork, with paintings selling for tens of thousands of dollars. Her work sold well in part because she was a child; this "pint-sized Pollock" had her first solo gallery performance at age four. Her paintings might be physically indistinguishable from those of others, but, as a child, Olmstead was untutored and isolated from the art world, and her work had the gloss of creative genius. The twist here is that her fame brought her to the attention of a television program called *60 Minutes II*, which profiled Olmstead, and later showed footage suggesting that her father was coaching her. This changed people's impression of the nature of her performance, and the value of her paintings plummeted.

PERFORMANCE ANXIETY

Can this essentialist theory tell us about what is and is not art?

Probably not. There is no sharp line between art and nonart. As

Dutton has argued, there are several properties that art typically possesses, and there is no right answer when faced with instances that contain just some of these properties. Also, art is a strangely self-conscious domain, and once a theory of art becomes popular, some smart-ass artist will rush to falsify it. The prime example here is that of one of the most influential works of art in the last century—Marcel Duchamp's *Fountain*, which was made in part to mock the theory that art must be beautiful.

Still, if art is a performance, two facts follow.

1. Artwork is intentional.
2. Artwork is intended to have an audience.

Intention first. We can leave behind us footprints on sand, crumpled paper in trash cans, and unmade beds, and none of these are usually art. But they can be art if they are made with the right sort of intention, and creations of this sort can be found in museums. Tracey Emin's *My Bed*, for instance, is her unmade bed with various objects on it; it was exhibited in the Tate Gallery. One can have two identical creations, then, but one gets to be art and the other does not based on the psychological states of their creators.

These are the intuitions of some philosophers, anyway. The psychologist Susan Gelman and I were interested in how well this matched our commonsense notion. So we tested three-year-olds by showing them objects and telling stories about their origin. We showed the children a blob of paint on a canvas, for instance, and either said that it was created by a child who accidently spilled his paint or by a child who used his paint very carefully. As predicted, this made a difference: When told that it was an accidental creation, children tended to later describe it using words like "paint";

but when told that it was created on purpose, children tended to describe it as art—as "a painting."

This brings us to the puzzle of how we should think about the creations of nonhumans, such as the paintings of certain elephants and chimpanzees. Many of these are very attractive, but it is hard to see them as art. The problem is that animals don't know what they are doing. If I dip my hamster's paws into paint and let her run across a canvas, it might look pretty, but it's not a painting. I doubt that the accomplishments of elephants and chimpanzees go beyond this. They don't plan their work and they don't admire it when they're done. The trained animals require human help, not just in the obvious way that they have to be given the tools, but also because they have to be stopped—if nobody pulls away the paint, the animals just keep going, creating canvases of brown smudge. What these animals do is strikingly different from the behavior of young children, who initiate the creation of art, stop when finished, and admire their work and show it to others.

This brings us to the second consideration, which is that art is meant to be displayed; it is created for an audience. This is what distinguishes it from other intentional activities like running a race, making coffee, combing your hair, or checking your e-mail. It is the difference between *Fountain* and a urinal; between Andy Warhol's *Brillo Box* and Brillo boxes; between John Cage's *4'33"* and some guy sitting at a piano for 4 minutes, 33 seconds because he is having a panic attack.

I admit that there are some counterexamples to this theory. There are creations that were never intended to be displayed—such as Rodin's sketches—but these are nonetheless thought of as artwork. Then there are objects that are intentionally created to be shown to an audience, but nobody would call them art. (*How Pleasure Works* was intentionally created for an audience, but it is not, in the usual sense of the word, artwork.)

Still, one of the virtues of the performance theory, with its emphasis on intention and the emphasis on audience, is that it includes so many of the right things. The claim that we intuitively see artwork as performances can give us some insight into how people can make sense of art that is unusual for its time, like the work of Duchamp, Warhol, and Cage. And it helps us make sense of our reaction to even more controversial artwork.

Take, for example, the Yale senior whose final art project purportedly involved repeated self-induced miscarriages. As she described it, early in her menstrual cycle, she would inseminate herself with the sperm of volunteers. On the 28th day of her cycle she would ingest an abortifacient and would then experience cramps and bleeding. The blood was part of the exhibit, along with video recordings. Her project generated controversy when it got picked up in the national news, and there was local debate over whether she really did all of these things, with the Yale administration insisting that it was a fraud.

Or consider the most famous work of Piero Manzoni: a series of 90 cans of the artist's feces. They sold well; in 2002, the Tate Gallery paid $61,000 for one of these cans. This is interesting artwork in many ways and nicely connects to the theme of essentialism. It is the epitome of positive contagion, the idea that the pleasure we get from certain objects is due to the belief that they contain a residue of the creator or the user. As Manzoni put it, "If collectors really want something intimate, really personal to the artist, there's the artist's own shit." Also, it comes with a wonderfully comic vision. Manzoni intentionally failed to properly autoclave the cans, and so at least half of these cans of feces, proudly on display in museums and private collections, later exploded.

People react to such cases in different ways. Many see them as shocking and ridiculous, while others are dazzled, getting great pleasure from them. My goal here is not to make a case pro or con, but to point out that even the harshest critic can appreciate that these are creative performances of a sort. We understand why the miscarriages took place (if they did), or why the artist is putting his poop into cans.

Furthermore, just as with Pollock, our clashing intuitions about the quality of such artwork derive in large part from what we think of the performance. If you have a low opinion of the capacities underlying a creation, then you will see it as bad art and get no pleasure from it—except, perhaps, the joy of mockery. We don't react to Manzoni's creation as we do to a Rembrandt (though, based on a discussion on the Tate Gallery Web site, it turns out to be surprisingly difficult to store poop in a can). You need to be dazzled by the idea to be dazzled by the art.

This is why people react so negatively to modern and postmodern work; the skill is not apparent. As the critic Louis Menand explains, artistic interest has shifted from the what of art to the how of art. Traditional art is about what is in the world; more modern works are about the very process of representation. An appreciation of much of modern art therefore requires specific expertise. Any dope can marvel at a Rembrandt, but only an elite few can make any sense of a work such as Sherrie Levine's *Fountain/After Marcel Duchamp*, and so only an elite few are going to enjoy it. Manzoni once flipped a pedestal upside down, so that its surface touched the ground, and then declared that the entire planet thereafter became his artwork. When I read about this, I found it pretty funny, but for me it's the sort of joke that a 10-year-old might make. Someone immersed in the art world probably sees it differently.

The play *Art* is a commentary on this tension between novices and experts. The play begins when Serge buys an unframed white canvas with some hard-to-see diagonal scars, and shows it to his friend Marc:

> *Marc*: You paid two hundred thousand francs for this shit?
>
> *[Later, Serge complains to another friend.]*
>
> *Serge*: I don't blame him for not responding to this painting, he hasn't the training, there's a whole apprenticeship you have to go through.

We don't have to give Serge the last word. Maybe Marc is right when he later insists that Serge sees something in the painting that does not exist. Marc would adore those real-life stories where the experts get it wrong, such as when David Hensel submitted his sculpture, a laughing head called *One Day Closer to Paradise*, to an open-submission contemporary art exhibition at the Royal Academy in London. He boxed it up with its plinth, a slate slab, for the head to rest on. The judges thought that these were two independent submissions, and they rejected the head but accepted the plinth. Expert intuitions about history and performance are not always accurate.

. . .

SPORT

We started with a musician in a subway station, moved to a famous forgery, and then turned to art more generally. We discussed many sources of artistic pleasure, but the focus has been on performance, the idea being that the pleasure we get from an artwork derives in part from our beliefs about how it was created.

This type of pleasure is not special to art. The Greeks were said to put sports and arts in the same category, something unheard of in modern scholarship. There are few university classes that study sport using the same intellectual tools used for art; there are professors who specialize in baroque music or pop art, but none who are experts on pole vaulting or soccer.

The dismissal of sport might be a mistake. Admittedly, art and sport differ in some obvious ways. Art is nonutilitarian, often aggressively useless. Sport has a more practical foundation; part of the pleasure of sport is the pleasure of practicing skills that were useful in the environment in which we evolved, such as running and fighting, and so humans might be motivated toward sport even if display weren't an issue. Perhaps because of this, art usually involves the expectation of an audience, but sport doesn't—if you play squash with a friend and nobody watches, it's still a game of squash.

Still, they are both displays of deeper human traits, and so the parallels run deep. For both, our appreciation of the performance is influenced by where and when it takes place. If I were to sign a urinal and submit it to a competition, I would not win. It's too late; Duchamp already did it in 1917. As one art expert writes: "To create something new is an achievement. Einstein was the first to see that $E = MC^2$. Afterward any actor could don a fuzzy wig and scribble the identical formula on a blackboard. That wouldn't make him Einstein."

Priority matters for sport as well. Some of this is also for reasons of originality, as with Muhammad Ali's audacious rope-a-dope against George Foreman in 1974. But for sport, priority matters for other reasons. When Roger Bannister broke the four-minute mile in 1954, this wasn't a creative act—it's not that people had never *thought* about running that fast—but it was nonetheless a unique and important performance. What made it special? Recall that Dutton writes about how our assessment of art is sensitive to "the ways in which artists solve problems, overcome obstacles, make do with available materials." Well, this is true for sport as well. Bannister had no coach; he trained with friends during lunch breaks while he was in medical school. Now any serious contender for a record in the mile would have access to a doctor, a coach, a nutritionist, and a masseur. It would be a full-time job, not something to be squeezed in around other commitments. We admire Bannister in part because his four-minute mile really was a performance superior to those who followed him.

Because art and sport are performances, there is the possibility of cheating in both. Cheating is intentionally misrepresenting the nature of one's performance. The standard example of artistic cheating is forgery, but there are other ways to cheat. We are dazzled by the speed of the double-jumps in a recorded performance of Liszt's "Mephisto Waltz"—until we discover that this was accomplished by a recording engineer. Crowds cheer at a live musical show, and then boo when they discover that it is lip-synched (as happened with Milli Vanilli in 1989). Cheating is obviously an issue for sport. Sports fans were no longer impressed by

Rosie Ruiz's feat of completing the Boston Marathon in less than 2 hours, 32 minutes in 1980 when it was discovered that she took the subway. It takes away from a pitcher's performance to know that he used spitballs or a boxer's victory to know that he had Plaster of Paris in his hand wrappings.

Then there is the moral outrage that arises with performances that are seen as artificially enhanced, through steroids and other drugs. As the writer Malcolm Gladwell puts it, steroids are seen as violating the "honesty of effort." An athlete on steroids is no longer credited as the author of the performance. But what makes steroids so much worse than acceptable interventions such as vitamins or weight machines or expensive swimsuits? My graduate student Izzat Jarudi interviewed Americans from New Haven and New York about the morality of steroids, and found that they strongly disapprove of them. What's interesting is that they were unable to explain why. Some mentioned worries about negative health effects, but when they were asked about steroids that were perfectly healthy, they still typically insisted that they should be illegal and that an athlete who uses them is a cheat.

Perhaps there is no rationale behind this intuition at all. These sorts of gut feelings are notoriously malleable. In vitro fertilization was shocking when first introduced; now it is something only a crank would object to. It is likely that much of what shocks us now will be commonplace in the future and that this moral outrage associated with steroids is largely rooted in an instinctive conservatism, a fear of the new.

Also, as Gladwell points out, there is something perverse about the worry that such an enhancement might give someone an unfair advantage. What makes this any more unfair than the natural advantage of being born with genes that make you strong? It is true that we have a gut feeling that there is a real difference here. But this might be because we are drawn to value natural gifts, since these are the sorts of capacities that are passed on to one's children. We admire the natural and disdain the enhanced. We've seen this before in the domain of beauty; people prefer natural beauty to hair implants and plastic surgery. These preferences might make good Darwinian sense and are difficult to override. But this doesn't make them fair.

PERFORMANCE GETS UGLY

Art and sport stand out as the sorts of performances that are especially valued. There are social structures that support them: art schools and sports camps, *Rolling Stone* and *Sports Illustrated*, the Louvre and Yankee Stadium, separate sections of the daily newspaper. But the joy of performance—the pleasure of seeing it and the pleasure of doing it—is more general than this, and more primitive.

Developmental psychologists have long marveled at how children naturally point, wave, and grunt to draw attention to interesting things in their environment. This might seem like the simplest skill until you realize that no other species does this. By some accounts, this desire to share our thoughts is responsible for much of what makes us human, including language and our sophisticated culture.

There is something else that may be equally important—the

impulse to show off certain skills. A toddler does a somersault, piles up blocks that don't fall, and stands on one foot. These are skill displays. They are sometimes explicitly done for a parent's approval, but children will do them alone; there is a pleasure in private play.

Some performances develop a competitive flavor. Every human society has foot races and wrestling matches. Anything can be grounds for competition. One child belches, and then another, and soon there is a belching competition. One seven-year-old tells a story, another tries to top it (the birth of fiction). Adolescents sit in a circle, telling jokes in turn, feeding off the laughter of others (the birth of standup comedy). The competition can be against your past self, as when runners try to beat their previous times. (My next-door neighbor, the economist Ray C. Fair, is a marathon runner, and he compares his times with his calculations of how they should decline by age). Crossword puzzles and Sudoku are other instances in which we might try to excel even with nobody else around.

We are a perverse and creative species, and there is no limit to the range of performances we can invent. As an eight-year-old, I knew that I was never going to be the fastest boy alive, but I was a demon on a pogo stick and tried for months, unsuccessfully, for the world record of the most consecutive bounces. I knew what I had to beat because I owned the *Guinness Book of World Records*, which is, as Dutton notes, a marvelous demonstration of all the ways in which humans can try to excel.

Not all performances are equal, in part because not all of them are fitness displays to the same extent. There is a pleasure to becoming expert at Sudoku, but it lacks the intellectual richness of chess.

One might marvel at the winner of The World Grilled Cheese Eating Competition (103-pound Sonya "The Black Widow" Thomas), but it's not quite the same as watching Rudolf Nureyev or Michael Jordan. Spelling bees are fine, but when selecting graduate students, I wouldn't be impressed with someone who was a national spelling champ. One can appreciate that there is a tremendous amount of discipline and coordination required to become the world champion at the video game *Donkey Kong*, but any pleasure in watching such a performance is tainted by the concern that the person is wasting his life.

Some displays have a paradoxical flavor. There has long been art that portrays ugliness, such as the paintings of Hieronymus Bosch. There is Duchamp's urinal, Manzoni's feces, Hirst's rotting cow's head, and countless contemporary works that utilize bodily fluids and animal parts. There is the story (probably apocryphal) of a sculpture by Ed Kienholz that had to be removed from the Louisiana Museum of Modern Art because people vomited when they saw it. One motivation for all of this ugly art is to refute the notion that art should be beautiful. Then there is the sense that beauty is too predictable, easy, accessible, and bourgeois. Bold and creative art must turn away from this. Many artists wouldn't be happy if you described their work as uplifting. There is also the appeal of the freak show; there is a perverse fascination in deformity, which is perhaps rooted in a less redeeming part of human nature, a drive toward sadism and mockery.

But sometimes ugliness can be more positive. In rural England, there are gurning competitions, where people compete to distort their faces into hideous positions. The rules are straightforward. Competitors put their heads through a horse collar and have a set time in which to contort their faces into the scariest or silliest expression possible. False teeth may be left in, taken out, or turned upside down if desired.

There is something impressive about this. Humans devote considerable energies to different forms of art, music, sport, and games, and, as I have argued here, these are typically displays of reproductively relevant capacities, the finest traits of humans: intelligence, creativity, strength, wit, and so on. We are essentialists, naturally drawn to the history of a performance, and so we can get pleasure from the display of such natural gifts. But we are also smart enough to turn this around and to occasionally get pleasure from the display of something that is, from a Darwinian perspective, exactly what we *don't* want. This is pleasingly egalitarian. Gurning is not yet an Olympic sport, but I hope one day it will be.

6

IMAGINATION

HOW DO AMERICANS SPEND THEIR LEISURE TIME? THE answer might surprise you. The most common voluntary activity is not eating, drinking alcohol, or taking drugs. It is not socializing with friends, participating in sports, or relaxing with the family. While people sometimes describe sex as their most pleasurable act, time-management studies find that the average American adult devotes just four minutes per day to sex—almost exactly the same time spent filling out tax forms for the government.

Our main leisure activity is, by a long shot, participating in experiences that we know are not real. When we are free to do whatever we want, we retreat to the imagination—to worlds created by others, as with books, movies, video games, and television (over four hours a day for the average American), or to worlds we ourselves create, as when daydreaming and fantasizing. While citizens of other countries might watch less television, studies in England and the rest of Europe find a similar obsession with the unreal.

This is a strange way for an animal to spend its days. Surely we would be better off pursuing more adaptive activities—eating and drinking and fornicating, establishing relationships, building shelter, and teaching our children. Instead, two-year-olds pretend to be lions, graduate students stay up all night playing video games, young parents hide from their offspring to read novels, and many men spend more time viewing Internet pornography than interacting with real women. One psychologist gets the puzzle exactly right when she states on her Web site: "I am interested in when and why individuals might choose to watch the television show *Friends* rather than spending time with actual friends."

One solution to this puzzle is that the pleasures of the imagination exist because they hijack mental systems that have evolved for real-world pleasure. We enjoy imaginative experiences because at some level we don't distinguish them from real ones. This is a powerful idea, one that I think is basically right, and I'll spend this chapter defending it and drawing out some of its more surprising implications. But I don't think this is *entirely* right, and the chapter that follows explores certain phenomena—including horror movies and masochistic daydreams—that require a different type of explanation, one that draws upon the same sort of essentialist theory proposed for food, sex, everyday objects, and art.

GREAT PRETENDERS

All normal children, everywhere, enjoy playing and pretending. There are cultural differences in the type and frequency of play. A child in New York might pretend to be an airplane; a hunter-gatherer child will not. In the 1950s, American children played

Cowboys and Indians; not so much anymore. In some cultures, play is encouraged; in others, children have to sneak off to do it. But it is always there. Failure to play and pretend is a sign of a neurological problem, one of the early symptoms of autism.

Developmental psychologists have long been interested in children's appreciation of the distinction between pretense and reality. We know that children who have reached their fourth birthday tend to have a relatively sophisticated understanding, because when we ask them straight out about what is real and what is pretend, they tend to get it right.

What about younger children? Two-year-olds pretend to be animals and airplanes, and can understand when other people do the same thing. A child sees her father roaring and prowling like a lion, and might run away, but she doesn't act as though she thinks her father is actually a lion; if she believed that, she would be *terrified*. The pleasure children get from such activities would be impossible to explain if they didn't have a reasonably sophisticated understanding that the pretend is not real.

It is an open question how early this understanding emerges, and there is some intriguing experimental work exploring this. My own hunch is that even babies have some limited grasp of pretense, and you can see this from casual interaction. A useful way to spend time with a one-year-old is to put your face up close and wait for the baby to grab at your glasses or nose or hair. Once there is contact, you pull your head back and roar in mock rage. The first time you get a bit of surprise, maybe concern, a dash of fear, but then you put your head back and wait for the baby to try again. It will, and then you give the pretend-startled response. Many babies come to find this hilarious. (If the baby is an eye-poker, you can wrestle over keys instead.) For this to work, though, the baby must know that you are not even a little bit angry; the baby must know that you are pretending.

They aren't perfect at this, of course. It is sometimes hard for anyone to tell the difference between goofing around and being serious, and you shouldn't expect too much from a creature the size of a Russian novel. Charles Darwin tells this story of his first son, William: "When this child was about four months old, I made in his presence many odd noises and strange grimaces, and tried to look savage; but the noises, if not too loud, as well as the grimaces, were all taken as good jokes; and I attributed this at the time to their being preceded or accompanied by smiles." But then William was fooled by his nurse: "When a few days over six months old, his nurse pretended to cry, and I saw that his face instantly assumed a melancholy expression, with the corners of his mouth strongly depressed."

Are play and pretense uniquely human? Dogs and wolves interact with one another in a way that looks like play, particularly play-fighting, and can even signal to one another that their attacks are not sincere, through "play bows" in which the animal crouches on its forelimbs, remains standing on its hind legs, and keeps its head lower than the animal that it is interacting with. This means, roughly, "I want to play" or "We're still playing." In a loose sense, this might count as pretend. But this sort of play is likely to be something that the animals are hardwired to do, as a means to practice important skills for later in life. They need not mentally encode it, at any level, as an imaginary version of real fighting.

Now, this is sometimes true for humans as well. When a child and a dog run through a park together, they might both be thinking the same thing—that is, not much of anything. But children can be smarter than this. There is a flexibility to their imaginations—anything real can be treated as pretense. You can show a child an entirely new action, such as cutting a paper in half, and then demonstrate a pretend version of the action (fingers as scissors, snipping at empty space) and, if you do it well enough, the child will get the idea—you are pretending to cut paper. This might seem simple, but I doubt that any animal other than a human can appreciate it.

METAREPRESENTATION

It is a special power, to hold something in one's mind, to reason about it and respond emotionally to it, but to know that it is not real. It shows the capacity to deal with metarepresentations—that is, representations about representations.

To get a sense of what this means, consider first the simplest thoughts that we have, such as

The umbrella is in the closet.

Such statements (or propositions) explain human action. If it is raining and you don't want to get wet, you might go to the closet for your umbrella, and this is because you believe that the umbrella is in the closet—you have something in your head corresponding to the sentence above. Other animals can do something similar. Rats, for instance, can encode propositions such as

The food is next to the corner.

Now consider the special part. Mary says that she doesn't want to get wet, and that she wants an umbrella, so she walks to the closet. You watch this and generate the belief:

Mary thinks the umbrella is in the closet.

This is a special sort of thought, because you can hold it in mind while believing that the embedded sentence is false—it's perfectly possible to believe that Mary thinks the umbrella is in the closet *and* to believe that the umbrella is not in the closet; it is actually in the living room.

This capacity to reason about another's false belief is important. It makes it possible to teach, a skill that involves keeping in mind that another person knows less than you do. It underlies lying and deception: when I say that I never got your e-mail even though I did get it, I am trying to put in your head a belief that is not true. Reasoning about false belief is difficult for children, though there are some recent demonstrations showing that if the task is made simple enough, even one-year-olds can succeed at it.

Metarepresentation is central to imaginary pleasure. We know, watching the play, that Jocasta is Oedipus's mother; what makes it a good story is that we also know that Jocasta and Oedipus do not themselves know this. The literary scholar and cognitive scientist Lisa Zunshine writes about a *Friends* episode in which Phoebe discovers that Monica and Chandler are romantically involved, and decides, as a joke, to flirt with Chandler. Monica discovers that Phoebe knows, so, in retaliation, she tells Chandler to welcome Phoebe's advances, so that Phoebe will have to back down, embarrassing herself, but Phoebe then realizes what Monica has in mind. She tells her friends: "They thought they could mess with us. They're trying to mess with *us*? They don't know that we know that they know we know!"

And Zunshine also gives the example of my favorite *New Yorker* cartoon:

"Of course I care about how you imagined I thought you perceived I wanted you to feel."

Where did this capacity for metarepresentation come from? There are two plausible, and compatible, accounts of the origin.

The first is illustrated in the examples above. Other people's actions are driven not by how the world really is, but by how they think the world is, and making sense of their behavior requires that you reason about facts that you know are not true. Metarepresentation might first evolve, then, in the context of understanding other minds.

The second possibility is that the capacity to imagine the unreal allows us to plan for the future, to evaluate worlds that do not yet exist and which may never exist. As the critic A.D. Nuttall put it, "I think the cleverest thing that Karl Popper ever said was his remark that our hypotheses 'die in our stead.' The human race has found a way, if not to abolish, then to defer and diminish the Darwinian treadmill of death. We send our hypotheses ahead, an expendable army, and watch them fall."

To see this at work, suppose you are planning a vacation. You might consider going to Ko Samet, an island in Thailand. You think of this and draw various conclusions based on what you know of this place—you conclude, for instance, that you will be close to a beach. It sounds like a fun place to visit. You compare this to the charms of spending a week in London, where you can walk to some excellent museums. Critically, these conclusions are segregated from one another and from current reality; they are of the form:

If I go to Ko Samet, I can walk to the beach.

If I go to London, I can walk to some excellent museums.

You have these beliefs without believing that the embedded propositions are true; you have them without believing that it is true, *right now*, that

I can walk to the beach.

I can walk to some excellent museums.

This might all seem obvious, but the capacity to create these segregated worlds allows us to plan in ways that no other creature can, because we can imagine and rank alternative futures. This is often rapid and unconscious, as when you turn down a second martini because you need to do work later in the evening and don't want to be groggy:

If I have the second martini, I will be groggy.

But it can be more deliberate—recall the discussion in Chapter 3 of Charles Darwin's ranking of the pros and cons of marrying Emma Wedgwood.

Both of these theories of the origin of metarepresentation are adaptationist. Once we have this system in place, though, our imaginative powers can be used for purposes that have no adaptive benefit, such as daydreaming, moviegoing, and reading.

Metarepresentation is central to pretend play. In an elegant study, the psychologist Alan Leslie got two-year-olds to pretend to pour water into a cup, and then to turn the cup over onto a bear. He found that they know the bear is actually dry, but they also know, in the pretend world, that the bear is soaked and needs to get wiped off, because

In this game, the cup is full.

And they know that it is true in pretense (just like in reality) that if you dump a full cup onto someone he or she gets soaked. My three-year-old niece points her finger at me and says, "Bang!" and I crumple to the ground, tongue lolling out, dead, but also, she knows, really alive.

STORY TIME

This "Bang!" example shows that imaginative pleasures don't have to be complicated. But they often are; they often come in the form of stories.

A promising perspective on stories is akin to how Noam Chomsky and his colleagues have described language, where differences are explained in terms of constrained variation on universal principles. For language, the universals have to do with certain aspects of meaning and specified ways to convey this meaning. For stories, they are universal plots.

I discussed a specific example of this in an earlier chapter—sexual subterfuge has long fascinated people, and the notion of someone pretending to be someone else in bed shows up in story after story, from ancient Hindu texts to the Hebrew Bible to episodes of *Buffy the Vampire Slayer.* Stories with bedtricks export naturally from one culture to another; the title of the movie *The Crying Game* was reportedly translated into Chinese in such a way that unfortunately gives away the main plot twist: *Oh No! My Girlfriend Has a Penis*. Good stories have universal appeal. While the particulars of *The Sopranos* would be impossible to follow by anyone from a sufficiently different culture (for instance, ironic references to the portrayal of Italian Americans on network television), the themes—worries about children, conflicts with one's friends, the consequences of betrayal—are universal.

The novelist Ian McEwan takes this universality claim further, proposing that you can find all the themes of the English nineteenth-century novel in the lives of pygmy chimpanzees: "alliances made and broken, individuals rising while others fall, plots hatched, revenge, gratitude, injured pride, successful and unsuccessful courtship, bereavement and mourning."

It is easy to miss this universality. McEwan points out that critics and artists in every generation will insist that they are doing something that nobody else has ever done before. After all, once we stop thinking like philosophers or scientists, it is the differences that matter. If I ask someone for directions in Seoul and he doesn't understand me, it's little solace to realize that, for a linguist, English and Korean are variants of the same universal language. If I'm choosing a novel at a bookstore, it is irrelevant that, at a sufficiently abstract level, all stories are the same. William James once quoted with approval "an unlearned carpenter," who said, "There is very little difference between one man and another; but what little there is, *is very important.*"

We shouldn't push the analogy between stories and language too far. For many linguists, the universality of language is because of a dedicated language organ or module. But there is no story organ or story module. Stories are similar because people have similar interests. The popularity of themes having to do with sex and family and betrayal, for instance, is not due to some special feature of the imagination, but rather because people are obsessed, in the real world, with sex and family and betrayal.

MOVED

It is often useful to think about the world as it isn't, but we haven't explained why we enjoy doing so. Isn't it odd that we are moved by stories, that we have feelings about characters and events that we know do not exist? As the title of a classic philosophy article put it, how can we be moved by the fate of Anna Karenina?

The emotions triggered by fiction are very real. When Charles Dickens wrote about the death of Little Nell in the 1840s, people wept—and I'm sure that the death of characters in J. K. Rowling's *Harry Potter* series led to similar tears. (After her final book was published, Rowling appeared in interviews and told about the letters she had gotten, not all of them from children, begging her to spare the lives of beloved characters such as Hagrid, Hermione, Ron, and, of course, Harry Potter himself.) A friend of mine told me that he can't remember hating anyone the way he hated one of the characters in the movie *Trainspotting*, and there are many people who can't bear to experience certain fictions because the emotions are too intense. I have my own difficulty with movies in which the suffering of the characters is too real, and many find it difficult to watch comedies that rely too heavily on embarrassment; the vicarious reaction to this is too unpleasant.

These emotional responses are typically muted compared to the real thing. Watching a movie in which someone is eaten by a shark is less intense than watching someone really being eaten by a shark. But at every level—physiological, neurological, psychological—the emotions are real, not pretend.

So real, in fact, that psychologists use fictional experiences to study and manipulate real emotions. If an experimental psychologist wants to see whether a sad mood helps or hurts people's ability to do logical reasoning (not a bad question, by the way), it is necessary to put the subjects in a sad mood. But to do so, the psychologist doesn't have to mess up their actual lives. Rather, the psychologist can show them a movie clip—such as the scene in *Terms of Endearment* where the character played by Debra Winger, in a hospital bed dying of cancer, sees her children for the last time. If someone comes to a clinical psychologist with a snake phobia, the first step isn't to get the client to cope with his fears by throwing a snake in his lap. Rather, one starts by getting the client to *imagine* the feared object; then the therapist can slowly ratchet it up to the real thing. This only makes sense if the response to the imagined snake and the response to the real snake correspond to points on the same scale—they both count as fear.

If the emotions are real, does this suggest that people believe, at some level, that the events are real? Do we sometimes think that fictional characters actually exist and fictional events actually occur? Of course, people do get fooled, as when parents tell their children about Santa Claus, the Tooth Fairy, and the Easter Bunny, or when an adult mistakes a fictional film for a documentary or vice versa. But the idea here is more interesting than that; it is that even once we consciously know something is fictional, there is a part of us that believes it's real.

It can be devilishly hard to pull apart fiction from reality. There are several studies showing that reading a fact in a story increases the likelihood that you believe the fact to be true. And this makes sense, because stories *are* mostly true. If you were to read a novel that takes place in London toward the end of the 1980s, you would learn a lot about how people in that time and place talk to one another, what they eat, how they swear, and so on, because any decent storyteller has to include these truths as a backdrop for the story. The average person's knowledge of law firms, emergency rooms, police departments, prisons, submarines, and mob hits is not rooted in real experience or nonfictional reports. It is based on stories. Someone who watched cop shows on television would absorb many truths about contemporary police work ("You have the right to remain silent . . ."), and a watcher of a realistic movie such as *Zodiac* would learn more. Indeed, many people seek out certain types of fiction (historical novels, for example) because they want a painless way of learning about reality.

We go too far sometimes. Fantasy can be confounded with reality; the publication of *The Da Vinci Code* led to a booming tourist industry in Scotland, by people accepting the novel's claims about the location of the Holy Grail. Then there is the special problem of confusing actors with the characters they play. Leonard Nimoy, an actor born in Boston to Yiddish-speaking Russian immigrants, was frequently confused with his best-known role, Mr. Spock from the planet Vulcan. This was sufficiently frustrating that he published a book called *I Am Not Spock* (and then, 20 years later, published *I Am Spock*). Or consider the actor Robert Young who starred in one of the first medical programs, *Marcus Welby, M.D.*, and who reported getting thousands of letters asking for medical advice. He later exploited this confusion by appearing in his doctor persona (wearing a white lab coat) on television commercials for aspirin and decaffeinated coffee. There is, then, an occasional blurring between fact and reality.

In the end, though, those who were brought to tears by Anna Karenina were perfectly aware that she is a character in a novel; those people who wailed when Rowling killed off Dobby the House Elf knew full well that he doesn't exist. And, as I mentioned earlier, even young children appreciate the distinction between reality and fiction. When you ask them, "Is such-and-so real or make-believe?" they get it right.

Why, then, are we so moved by stories?

ALIEF

David Hume tells the story of a man who is hung out of a high tower in a cage of iron. He knows himself to be perfectly secure, but, still, he "cannot forebear trembling." Montaigne gives a similar example, observing that if you put a sage on the edge of a precipice, "he must shudder like a child." My colleague, the philosopher Tamar Gendler, describes the Grand Canyon Skywalk, a glass walkway that extends 70 feet from the canyon's western rim. Standing on it is a thrilling experience. So thrilling that some people drive several miles over a dirt road to get there, and then discover that they are too afraid to step onto the walkway. In all of these cases, people know they are perfectly safe but they are nonetheless frightened.

In an important pair of papers, Gendler introduces a novel term to describe the mental state that underlies these reactions: she calls it "alief." Beliefs are attitudes that we hold in response to how things *are*. Aliefs are more primitive. They are responses to how things *seem*. People in the above examples have beliefs that tell them they are safe, but they have aliefs that tell them they are in danger. Or consider Paul Rozin's findings that people often refuse to drink soup from a brand-new bedpan, eat fudge shaped like feces, or put an empty gun to their head and pull the trigger. Gendler notes that the beliefs here are: the bedpan is clean, the fudge is fudge, the gun is empty. But the aliefs are stupider, screaming, "Dangerous object! Stay away!"

The point of alief is to capture the fact that our minds are partially indifferent to the contrast between events that we believe to be real versus those that seem to be real or that are imagined to be real. This extends naturally to the pleasures of the imagination. Those who get pleasure voyeuristically watching real people have sex will enjoy watching actors having sex in a movie. Those who like observing clever people interact in the real world will get the same pleasure observing actors pretend to be such people on television. Imagination is Reality Lite—a useful substitute when the real pleasure is inaccessible, too risky, or too much work.

Humans have invented many ways of exploiting alief, of creating surrogates of pleasant real-world experiences. We can do this with stories or even in wordless play (consider a parent swinging a child into the air, creating the sensation of flying). We can use the presence of actors on a stage or screen as imaginative aids, narrowing the gap between real and virtual experiences. We can generate our own pleasant aliefs, in daydreams. If you would enjoy winning the World Series of Poker, flying around Metropolis, or making love to a certain someone, then you can get some limited taste of these pleasures by closing your eyes and imagining these experiences.

(This might not seem like much of a trick, but I doubt that any other animal has come across it. Dogs dream, but do they daydream? My dog spends most hours doing nothing; she is next to me now as I write this, staring at me. When people are left alone, they plan, daydream, and fantasize, but I'm not sure if Tessie does any of this. I don't know what—if anything—is going on in her head. The same question can be asked about closer evolutionary neighbors: Monkeys are famously enthusiastic masturbators, for instance, but do they have sexual fantasies while they do it? Was the writer Lin Yutang right when he suggested "the difference between man and the monkeys is that the monkeys are merely bored, while man has boredom plus imagination"?)

Often we experience ourselves as the agent, the main character, of an imaginary event. To use a term favored by psychologists who work in this area, we get *transported*. This is how daydreams and fantasies typically work; you imagine winning the prize, not watching yourself winning the prize. Certain video games work this way as well: they establish the illusion of running around shooting aliens, or doing tricks on a skateboard, through visual stimulation that fools a part of you into thinking—or alieving—

that you, yourself, are moving through space. Psychological studies suggest that this is a natural default when reading a story; you experience the story as if you are in the character's head.

For stories, though, you have access to information that the character lacks. The philosopher Noël Carroll gives the example of the opening scene in *Jaws*. You can't be merely taking the teenager's perspective as she swims in the dark, because she is cheerful and you are terrified. You know things that she doesn't. You hear the famous, ominous music; she doesn't. You know that she is in a movie in which sharks eat people; she thinks that she is living a normal life.

This is how empathy works in real life. You would feel the same way seeing someone happily swim while a shark approaches her. In both fiction and reality, then, you simultaneously make sense of the situation from both the character's perspective and from your own.

GOOD FOR NOTHING

This approach can explain the general appeal of stories. Stories are about people, and we are interested in people and how they act. It is not hard to imagine an evolutionary purpose for why we would care about the social universe; indeed, it's been argued that one main force in the evolution of human language is that it is a uniquely powerful tool for communication of social information—and, particularly, gossip.

Not all imagined worlds include people—as I write this, a book was just published called *The World Without Us*, which provides an imaginative re-creation of what Earth would be like if humans became extinct. Such purely nonsocial worlds are the exception, though, and many of the cases that look at first to be nonsocial are really about people—popular science books and documentaries, at least the successful ones, are often about the scientists themselves, their histories, personal clashes, and so on. Lisa Zunshine makes a similar point when she notes that purely nonsocial descriptions of nature are scarce in novels, even in novels that have the reputation for this sort of thing. "It is possible," she suggests, "that our perception of some fictional texts as abounding in such descriptions owes simply to the fact that relatively rare as they are, they stand out." When such descriptions do occur, as in the novels of Ivan Turgenev, they bustle with intention; they exhibit the *pathetic fallacy*—John Ruskin's term for the act of imbuing natural objects with thoughts, sensations, and emotions.

Our interest in people motivates some quirky pleasures. For most of the history of the species, the goings-on of important people really mattered. These people held sway over our lives, we needed to learn from them, curry their favor, avoid their wrath, and so on. As we find ourselves in societies with thousands of people, then millions, then billions, this obsession persists. The death of Princess Diana, for instance, was a profoundly moving event for much of the world, as was the 2005 breakup of the actors Brad Pitt and Jennifer Aniston. We have a hunger for social information, and celebrity gossip and fictional stories sate us with irrelevant tales of people who don't matter and people who don't exist. It is as if we are starving to death, and we gorge ourselves on calorie-free sugar substitutes.

Is this it? Is the pleasure of fiction just an accident, a by-product of the fact that our emotions don't care about whether an event or person is real or make-believe?

This is too minimal for some, and many scholars seek out an

adaptive account of the pleasure of stories. Zunshine argues that we are driven to enjoy stories because they serve to exercise our social capacities; they give us useful practice in thinking about the minds of other people. The psychologists Raymond Mar and Keith Oatley suggest that the function of fiction is to acquire social expertise. Denis Dutton and Steven Pinker explore variants of the claim that fiction helps us explore and learn about solutions to real-world dilemmas. As Pinker puts it: "The cliché that life imitates art is true because the function of some kinds of art is for life to imitate it."

I don't doubt that stories can do all of this and more. They can also instill moral values and inspire moral change—elsewhere, following scholars such as the philosopher Martha Nussbaum, I have argued that stories are a primary mechanism for how societies get nicer, how novel moral insights such as the evil of slavery can be packaged in a form that persuades others and eventually becomes accepted as the status quo. Stories can also alleviate loneliness. They can help you win friends and attract potential mates—being a skilled raconteur is an attractive quality. And, as I will discuss in the next chapter, stories can provide a mechanism for safe practice, an arena through which to mentally prepare for certain unpleasant situations.

Stories can do all of these things. But this is not why we have them. As evolutionary explanations, such accounts are superfluous. Once you have a creature that responds with pleasure to certain real-world experiences and doesn't fully distinguish reality from imagination, the capacity to get pleasure from stories comes for free, as a lucky accident.

• • •

SAD AND POWERFUL

In his *Introduction to Shakespeare*, Samuel Johnson writes: "The delight of tragedy proceeds from our consciousness of fiction; if we thought murders and treasons real, they would please no more."

Samuel Johnson was a brilliant writer, but plainly he had never heard of O. J. Simpson. If he had, he'd realize that we get plenty of pleasure from real tragedy. Indeed, Shakespeare's tragedies depict precisely the sorts of events that we most enjoy witnessing in the real world, complex and tense social interactions revolving around sex, love, family, wealth, and status.

Both for tragedy in particular and for negative events more generally, reality tends to be more interesting than fiction. When a memoir is discovered to be fictional, its sales go down, not up. In the last few decades, when a horrific event occurs—such as Susan Smith's drowning of her children or the random murders by the Washington, D.C., snipers—the immediate reaction is to make a movie about it. The plausible assumption here is that the reality of the event will pump up the interest.

I have argued that our emotions are partially insensitive to the contrast between real versus imaginary, but it is not as if we don't care—real events are typically more moving than their fictional counterparts. This is in part because real events can affect us in the real world (fictional snipers can't shoot those you love; real ones can), and in part because we tend to ruminate about the implications of real-world acts. When the movie is finished or the show is canceled, the characters are over and done with. It would be odd to worry about how Hamlet's friends are coping with his death because these friends don't exist; to think about them would involve creating a novel fiction. But every real event has a past and a future, and this can move us. It is easy enough to think about the families of those people whom O. J. Simpson was accused of murdering.

But there are also certain compelling features of the imagination. Just as artificial sweeteners can be sweeter than sugar, unreal events can be more moving than real ones. There are three reasons for this.

First, fictional people tend to be wittier and more clever than friends and family, and their adventures are usually much more interesting. I have contact with the lives of people around me, but these people tend to be professors, students, neighbors, and so on. This is a small slice of humanity, and perhaps not the most interesting slice. My real world doesn't include an emotionally wounded cop tracking down a serial killer, a hooker with a heart of gold, or a wisecracking vampire. As best I know, none of my friends has killed his father and married his mother. But I can meet all of those people in imaginary worlds.

Second, life just creeps along, with long spans where nothing much happens. The O. J. Simpson trial lasted *months*, and much of it was deadly dull. Stories solve this problem—as the critic Clive James once put it, "Fiction is life with the dull bits left out." This is one reason why *Friends* is more interesting than your friends.

Finally, the technologies of the imagination provide stimulation of a sort that is impossible to get in the real world. A novel can span birth to death and can show you how the person behaves in situations that you could never otherwise observe. In reality you can never truly know what a person is thinking; in a story, the writer can tell you.

Such psychic intimacy isn't limited to the written word. There are conventions in other artistic mediums that have been created for the same purpose. A character in a play might turn to the audience and begin a dramatic monologue that expresses what he or she is thinking. In a musical, the thoughts might be sung; on television and in the movies, a voice-over may be used. This is commonplace now, but it must have been a revelation when the technique was first invented, and I wonder what young children think when they come across this for the first time, when they hear someone else's thoughts expressed aloud. It must be thrilling.

As another case of intimacy, consider the close-up. Certainly voyeurism has long been a theme of movies, from *Rear Window* to *Disturbia*, but the technique of film itself offers a unique way to satisfy our curiosity about the minds of others. Where else can you look full into someone's face without having the person look back at you? "Some viewers thrill to the prospect of views into the bedroom and bathroom," the philosopher Colin McGinn writes, "but the film viewer can get even closer to the private world of his subject (or victim)—to his soul."

So while reality has its special allure, the imaginative techniques of books, plays, movies, and television have their own power. The good thing is that we do not have to choose. We can get the best of both worlds, by taking an event that people know is real and using the techniques of the imagination to transform it into an experience that is more interesting and powerful than the normal perception of reality could ever be. The best example of this is an art form that has been invented in my lifetime, one that is addictively powerful, as shown by the success of shows such as *The Real World*, *Survivor*, *The Amazing Race*, and *Fear Factor*. What could be better than reality television?

7

SAFETY AND PAIN

WOULD YOU ENJOY WATCHING A MOVIE OF HEAD SURGERY being performed on a young girl, starting with her face being pulled away from her skull? I doubt it. When the psychologist Jonathan Haidt and his colleagues showed this movie to undergraduates, they rated it as disturbing and disgusting, and few watched it until the end. A film of a monkey being beaten unconscious and its brain scooped out and served on plates got the same reaction.

The previous chapter explored a simple theory of imaginative pleasure: our minds are partially indifferent to whether an experience is real. If you would be aroused watching real sex, you can be aroused watching actors have sex; if you are interested in love and betrayal, you would be interested in a novel that describes love and betrayal. The pleasures of the imagination are parasitic on the pleasures of real life.

This can't be a complete theory, however. Sometimes what is

terrible or boring or depressing in reality is intensely enjoyable in the imagination. We enjoy fictions that make us cry, haunt our dreams, and gross us out. We do things in virtual worlds that would shock us in the real one, and our daydreams are not always pleasant ones; even happy people obsess on their worst fears. Here I try to explain why.

THE BEST STORIES

Insofar as stories are surrogates for real events, the best are those where we forget they are stories. Many writers aspire toward this. Elmore Leonard warns storytellers to avoid "hooptedoodle"—anything that calls attention to the writer and away from the story. Richard Wright wrote that he wanted to "fasten the reader upon words so firmly that he would forget words and be conscious only of his response."

Reading requires effort by the reader; it is easier to lose oneself in a movie. There are those who cringe in their seats during horror movies, peeking through fingers, and there is the old story of early filmgoers who screamed and dove for cover when the gun on the screen turned toward them and fired. A movie is the closest we have to virtual reality, and, as the philosopher Colin McGinn has stressed, it is experienced best on the big screen. You lose the force of the experience on a small television set or, worse, in the corner of a computer display, next to e-mail and the Web browser.

Technology might ultimately bring us to the point where the only difference between reality and fiction is going to be our explicit knowledge of which is which. Perhaps one day we can do away even with this explicit knowledge. People might pay for a virtual experience that, like a dream, is thought to be real while they are having it. Maybe this is what's happening to you right now. René Descartes worried that all of his experiences were sham, that he was being deceived by an evil demon. Perhaps we are brains in a vat or living in the Matrix. The philosopher Robert Nozick turns this worry into a pleasure technology, imagining a virtual-reality machine that can give one the illusion of living a life of immense pleasure, wiping out the memory of choosing to be in the machine. (Are you happy now? Perhaps you are in Nozick's machine.)

As a low-tech version of this idea, imagine that your friends hire actors to immerse you in a world that they have thought up, perhaps a thriller or a romantic comedy. You would be the main character in a story that you didn't know was a story. It would be disappointing when it ended, but while you were in the midst of it, it could be as exciting as life itself.

Still, something would be missing. Some of the pleasure that we get from books and movies and the like requires the appreciation that the imagined world is the intentional creation of others.

Take an example from art. Think about entering a strange house. You walk into the living room and see, through a window, a baby in diapers on the lawn, napping on a blanket. It is an attractive scene, and you walk closer to the window to get a better look, and then you see that it is not a window at all, but a sharply realistic painting, a *trompe l'oeil*. At this instant, a switch is thrown. As you study the painting, there is a thrill of appreciation—the baby was fine to look at, but you are really moved now by this amazing artwork. You have shifted to a different, powerful, source of appreciation.

Or imagine sitting on a plane eavesdropping on a whispered conversation by a couple in the seats behind you. It might be captivating (Did you try to kiss her? No, but I wanted to. *Bastard.*) or it might just go on, as conversations sometimes do (Did you pick up new lightbulbs? We have lightbulbs in the kitchen cabinet. No, we don't. *Yes, we do.*). Either way, it's not for you or for anyone else. It's something real in the world; its appeal is based on its intrinsic qualities.

But what if you discover that this was a form of street theater, that the couple was talking for your benefit? A switch is thrown, and you see it in a new way. The dialogue now has a point; you can be affected by it, impressed by its intelligence and its imagination, or disappointed by its predictability or crudeness. Such responses are different from how you were reacting when you thought it was real.

Fiction is a form of performance and, as such, we take pleasure from what we see as the virtuosity and intelligence of its creator. There is the thrill of being in the hands of someone who controls the story, someone who persuades and entrances and misdirects, someone who is (in this domain, at least) smarter than we are.

This is illustrated best in humor. Laughter is triggered by social events; the physical world is rarely amusing. If you see someone walking alone and laughing, he is either talking on a cell phone, remembering an amusing event, or schizophrenic. Laughter-worthy situations often arise because people think them up, as in the classic guy-slips-on-a-banana-peel scenario. The standard version starts with the man walking, cuts to the peel, cuts to a wide shot of the man approaching the peel, back to the peel, then his foot hits the peel, and then he falls. This can be funny, particularly if you haven't already seen it a thousand times and if the actor is skilled at conveying his surprise. But the same situation is not typically funny in real life. I spent much of my life in Montreal and I've seen many people tumble on ice on city streets. Onlookers wince or they reach to help or they turn away, but they typically don't laugh. This is funny in fiction, not in real life.

(I should qualify this: it probably would make people laugh to see someone slip on *a banana peel*, but that's just because it would be instinctively seen as an unintentional homage to slapstick comedy.)

Charlie Chaplin once described an improvement on the banana peel sequence. It starts with the guy walking, cuts to the peel, cuts to a wide shot of the guy approaching the peel, back to the peel, and then, when his foot is about to hit the peel, he steps over it—only to fall into an open manhole. This was Chaplin's version; I once saw a variant, in which the man steps over the peel and then is hit by a truck. Either way, it's hilarious, in part because of the appeal of fictional violent death (a topic that I'll turn to later), but mostly because the filmmaker has faked us out; our laughter is a form of applause. This is similar to the horror movie trick where there is a close-up camera shot of a teenager opening the closet door, ominous music, the generation of tension, then, boom—a loud noise, a sudden motion—but it's just the cat. And, again, people laugh, because they recognize the intelligence of the creation, they know that they have been intentionally misled. There is nothing like this in reality.

Admittedly, we sometimes ignore the fact that an imaginative world is an intentional object. One can "fall into" a movie or book, be transported, and therefore be blind to the extraordinary work required to give the appearance of reality and make the author or director disappear. The existence of the creator is often most salient when something goes wrong, when the costume looks inappropriate or the dialogue is unrealistic.

Still, even if we aren't consciously thinking about the fictional world as intentionally created, we are sensitive to the choices that the creator makes. We respond to these choices, and our responses are sensitive to our experiences with the genre. Laugh tracks were a clever discovery, exploiting the contagiousness of laughter to make television comedies funnier, but they have grown in recent years to be seen as cheap and manipulative, and many television shows now do without them. Critically, it is our reaction to laughter *in television shows* that has changed, not our reaction to laughter itself, just as over the last hundred years, the Western taste in portrait painting has changed, distinct from any change in how we look at real faces.

The writer A. J. Jacobs notes that in nineteenth-century France, theaters would hire *claques*, which included specialists in laughing, shouting for encores, and crying. Jacobs has the clever idea of adding canned crying to television shows—you'd be watching a medical show "and a softball player would come in with a bat splinter through his forehead, and you'd hear a little whimper in the background, turning into a wave of sobs." At first, this canned crying would be odd and distracting, but once it became commonplace, we'd be struck by a show that was edgy enough to do without it.

In his provocative book *Everything Bad Is Good for You*, science writer Steven Johnson makes a more general claim about how our expectations have changed. He points out that watching television shows from 20 or 30 years ago can be agonizing, with their painfully slow pace, simpleminded plots, and blaring laugh tracks. He contrasts this with modern shows like *24* in which multiple story lines are intertwined, dialogue tends to be more cryptic and realistic, and so on. Our tastes have changed. Johnson makes the provocative claim that this change corresponds to a rise in our intelligence, but I am tempted by a milder conclusion, which is that we have gotten smarter *about television*. We have developed the ability to cope with more from this medium, and our expertise has shaped our preferences.

When we enjoy a fiction, then, our aesthetic response is often a reaction to the creator's cleverness, knowledge, wit, and so on. As with performances in sport, music, and painting, these can provide pleasure.

There are other forms of pleasurable human connection. A child might enjoy hearing a story told by his mother, simply because of the intimacy of this connection. And some fictions infuse us with admiration, even awe, of the moral vision of the author. The literary scholar Joseph Carroll makes this point with a fictional example, that of Dickens's character David Copperfield who discovers a series of books that belonged to his dead father: "What David gets from these books is not just a bit of mental cheesecake, a chance for transient fantasy in which all his own wishes are fulfilled. What he gets is lively and powerful images of human life suffused with the feeling and understanding of the astonishingly capable and complex human beings who wrote them."

SAFE

Throwing the switch—going from seeing something as real to seeing something as an intentional creation—makes possible certain aesthetic pleasures. There is another effect as well. Once you know something to be fictional, you can expect your experience to be safe or at least safer than its parallel in reality.

What does *safe* mean? In part, literally safe. Someone observing a real barroom fight might take a beer bottle to the face; someone eavesdropping on a real conversation runs the risk of being caught and embarrassed. This is not a problem with books and movies. Indeed, as mentioned in the last chapter, one of the striking features of fiction is that one can safely observe people in their most intimate moments, pressing up to their faces and bodies in ways that one could never do in real life. It is only in fiction that you can look into the eyes of someone who is not looking back at you.

Also, in fiction, everyone else is safe as well. One can get upset when terrible things happen to fictional characters, because we respond to imagined events as if they were real—the problem of alief. But this upset is muted by the explicit knowledge that real men, women, and children are not affected. This blunts the empathetic cost of fiction.

Stories are safe in a third, subtler, sense. The real world just *is*; unless you believe that God's hand is everywhere, much of life has no point. If your phone wakes you up and it's a wrong number and you can't go back to sleep, that's bad luck. If you see a gun in the morning, it won't necessarily go off during the afternoon. But there are no accidents in stories. If you are watching a movie in which the phone wakes a character in the middle of night, it has significance. It was someone checking whether she was at home. It was her doppelganger! It was just a wrong number, but once awake, she walks to the bathroom and stares into the mirror and realizes that Bob never loved her. People spend much of their lives sleeping, checking their e-mail, going to the toilet, and watching television, but these activities are rarely shown in movies, because they are rarely relevant to the goals of the author. (Some more experimental filmmakers emphasize these uninteresting, or accidental, parts of life, but this is an intentional choice as well.) Our knowledge that everything in a story has a purpose shapes what we expect and what we like.

This particular type of predictability, sadly, takes away some of the pleasure of movies. There is such a thing as too safe. While chasing a beautiful Danish assassin over the rooftops of an Albanian slum, James Bond leaps from building to building. This might be good fun, but it's not *that* thrilling because anyone with any knowledge of movies knows that Bond is not going to fall. He is invulnerable. It would be quite the treat to see a James Bond movie in which, after the opening credits, he is chasing the assassin, runs toward the ledge, slips on a banana peel, and falls screaming onto the street below. Then, closing credits. This won't happen and our knowledge that it won't happen diminishes the pleasure of the scene. In this regard, children can get more pleasure from thrillers than adults, because they are less conscious of these conventions.

UNSAFE FOR CHILDREN

Just as with adults, hearing "once upon a time" throws a switch in children's minds. They distinguish fiction from reality. They know that Batman doesn't exist and that their best friend does. They know that the fantastic events in storybooks cannot happen in reality and that storybook creatures such as dragons need not follow normal rules of nature. They know that a real cookie can be touched and eaten but an imaginary cookie cannot, and they describe ghosts, monsters, and witches as "make-believe," as opposed to dogs, houses, and bears, which are "real-life."

In a series of experiments, the psychologist Deena Skolnick Weisberg and I found that preschool children go beyond this and understand, as adults do, that there are *multiple* fictional worlds. The motivation for our studies was the observation that for adults there is a complex commonsense cosmology of reality and imagination. There is the real world, but also a world for Batman and Robin, a different world for Hamlet, a third for the Sopranos, and so on. These worlds can interact in intricate ways—the world of Tony Soprano and his Mafia family makes contact with the world of Batman, for instance, but just in the sense that Tony, like us, thinks that Batman is a fictional character.

We find that four-year-olds grasp some of these complexities. They agree that Batman, Robin, and SpongeBob SquarePants are make-believe, and they understand that Batman thinks that Robin is real (because they are in the same world) and that Batman thinks that SpongeBob is make-believe (because they are in different worlds).

When it comes to the imagination, then, children are smart. But they are also vulnerable. The major problem has to do with alief—the fact that the mind doesn't fully care about the difference between what is known to be real versus known to be imagined. Nobody ever caught a stray bullet while reading about a gunfight, but we can be upset, even traumatized, by what we know is not real. In the last chapter, I discussed this from the standpoint of someone passively experiencing an imagined world, but it applies more forcefully to someone who is playing a fictional role. Consider going to someone you love and saying that you are going to do some acting. Explain that you will scream, "I hate you. I hope you die." Reassure them that this is an experiment and that the line is scripted. (Show them this page.) Still, I think it would be unpleasant to say and unpleasant to hear, and I don't recommend doing it. On a nicer note, though, it is sometimes said that actors who play lovers on the stage often fall in love for real. And therapists sometimes advise the depressed to act as if they are happy; smiling can have a positive effect on mood. Score one for the powers of alief.

All of this is more intense for children. It would be wicked to ask your five-year-old to play a game in which you pretend to hate him and scream at him that he is worthless. Children can understand that you are pretending, but they have a harder time than adults in blocking the emotional force of imagined experience. This pretend abuse would be real abuse.

In a milder demonstration of this, psychologists showed a box to young children and asked them to pretend that there was a monster inside it. When later given the chance to approach the box, the children often refused to put their fingers in the box. It's not that they really believed in the monster, it is that the imagined monster takes on such force in the child's mind that it is felt as if it were real. Children are too easily overwhelmed by the imagination. This is also why we shield them from certain fictions. You don't need to see a research study to understand that horror movies can give children nightmares.

Children are different in this regard from adults in degree, not in kind. My bet is that if adults were tested in the monster-box study, they would hesitate for a fraction of a second before putting their fingers in, in the same way that we don't like to eat a turd-shaped piece of fudge, or drink soup from a new bedpan, or sip water from a cup labeled "cyanide" even if we know that it is fresh from the tap.

Children are also vulnerable in that they know less about how stories work. In a series of experiments, Deena Skolnick Weisberg, David Sobel, Joshua Goodstein, and I gave preschool children the beginnings of stories and asked them to choose appropriate continuations. Some of the beginnings were realistic, such as a boy riding his scooter; some were fantastic, such as a boy who could make himself invisible. We expected children either to assume that realistic stories would continue in a realistic way and fantastic stories would continue in a fantastic way, as adults do, or to prefer fantastic continuations for all stories, following the idea that children are inclined toward magical thought. To our surprise, the children showed a bias *against* the fantastic; they preferred realistic continuations regardless of the story.

Children's ignorance makes the stories less safe. Several years ago, I was watching *Free Willy 2* with my family, and there is a scene in which the characters are on a sinking raft. My son Zachary, who was five, became agitated and started to whisper that they would drown. I explained that everyone would be fine. He asked me how I could know, since we hadn't seen the movie before, and I said that I knew how this sort of movie works—a feel-good family film will not kill the adorable children. I was right, and now he knows this too.

Later that year, we were canoeing in a river near our house, and we capsized. Panicked, Zachary shouted that we were going to drown. We had life jackets and the river was only three feet deep, but he wasn't being *unreasonable* here. Unless one believes that a divine being scripts our lives, reality lacks the constraints of fiction. Life is not a PG movie; sometimes the adorable children end up dead.

SADISTIC AND HORRIBLE

How does this safety transform fictional experience?

For one thing, it helps us take pleasure in the pain and death of others. You can laugh hard at the slapstick scene where the pedestrian plummets into the manhole, because you don't worry that he will be killed or crippled for life, you don't think about the grief of the wife and kids, you don't worry about any of this because you know the character doesn't really exist.

The increased tolerance for violence is manifest in the pleasure of video games. Often, they provide watered-down versions of real-world pleasurable experiences, as with flight simulators and racing games, which simulate the pleasures of flying and racing. Much of the violence in video games can be explained this way. In the typical game, you are immersed in a simulation in which one is doing something both exciting and morally good—defending the world from attacking aliens, killing Nazis, killing zombies, killing Nazi zombies—the sort of thing that, if it were safe, video game players would love to do in the real world.

But there is a darker pleasure too. The safety of video games allows people to exercise their worse impulses. Most players occasionally choose to shoot a teammate in the head, run over a civilian, or fly their plane into a building (in the case of *Microsoft Flight Simulator*, created in 1982, the easiest target was the Twin Towers in New York City). A while ago, while playing *The Sims*, a computer game where you create your own imaginary world, my children and I deprived a man of food and sleep for several days, and we watched as he screamed and begged and cried. When he died, we cheered.

This gets worse. In *Grand Theft Auto*, you can murder prostitutes. There are games, such as *Rape Lay*, imported from Japan, in which the commission of evil is the primary goal. One does wonder about someone who plays such games. In any case, the safety of these games (safe from harm, safe from the law, safe from worries about real people) allows for the expression of sadistic impulses that people would presumably not exercise in real life.

SAFETY MIGHT also help us solve a long-standing puzzle of fictional pleasure, one that was beautifully summarized by David Hume in 1757:

> *It seems an unaccountable pleasure which the spectators of a well-written tragedy receive from sorrow, terror, anxiety, and other passions that are in themselves disagreeable and uneasy. The more they are touched and affected, the more they are delighted with the spectacle. . . . They are pleased in proportion as they are afflicted, and never are so happy as when they employ tears, sobs, cries, to give vent to their sorrow, and relieve their heart, swoln with the tenderest sympathy and compassion.*

Hume is marveling at the fact that viewers of a tragedy get pleasure from emotions that are normally not good ones to have, such as sorrow, terror, and anxiety—the more of these emotions they get, the happier they are.

This puzzle comes into sharper relief when we turn to what philosopher Noël Carroll calls "the paradox of horror." Unlike tragedies, horror movies often have no redeeming aesthetic or intellectual qualities. But people like them, lining up to see innocents killed, tortured, and eaten by creatures such as zombies, axe-wielding psychopaths, sadistic aliens, swamp things, really mean babies, and, in one classic that I remember from way back (*Rabid*), a phallic growth coming out from an attractive woman's armpit. The last decade has brought us movies such as *Hostel* and the *Saw* series, in which the main point is the depiction of sadistic torture. These are not restricted to some pervy niche. You can find torture porn in the multiplex next to thoughtful dramas about divorced women finding love again and goofy comedies with smart-aleck donkey sidekicks.

Keep in mind that the puzzle is not just how we override the unpleasantness of death and pain. The problem is why we like it so much. *Friday the 13th* would not have been a more popular movie if Jason had attacked people with a Nerf baseball bat, just as *Hamlet* wouldn't have been a better play if he'd lived happily after ever. People enjoy scary movies *because* they are scary. At least at a visceral level, modern films are far scarier than those ever made in the past, and this reflects supply and demand. The more frightening the movie, the better. As Hume would have put it if he were around today: the negative emotions are not a bug; they are a feature.

The appeal of this sort of unpleasantness is not necessarily low-brow. In 2008, the *New York Times* had a discussion of *Blasted*—a very popular play, with sold-out performances and superb reviews. The article discussed a scene in which one man rapes another, and then sucks out his eyeballs and eats them. The audience for this play is older, sophisticated, and well-off; they are not hooting adolescent boys trying to out-macho one another. But nobody concerned with the production believes that it would be more popular if they ramped down the rape and cannibalism a bit. The audience loves the scene; it's one reason why the play is so popular.

There is a theory of what goes on here, which comes first from Aristotle but was elaborated and made famous by Freud. It is *catharsis*—certain events initiate a psychological purging process, through which fear and anxiety and sadness are released, and we feel better, calmer, and purified afterward. We suffer through the aversive experiences, then, because of the positive payoff at the end—for the release.

Perhaps this happens sometimes—there are people who claim to feel better after a good cry—but catharsis is a poor theory of the emotions, one that has no scientific support. It is just not true that emotional experiences have a purging effect. To take a much-studied case, watching a violent movie doesn't put one in a relaxed and pacifistic state of mind—it arouses the viewer. People don't leave horror movies feeling mellow and safe; they don't walk out of tragedies feeling giddy. The typical result of feeling bad is feeling worse, not feeling better. The pleasure of horror and tragedy, then, can't be explained as some sort of blissful afterglow.

PREPARE FOR THE WORST

Put aside fiction for a moment and consider another puzzle: Why do young animals, including young humans, play-fight? Why do children get pleasure from grappling and punching and knocking each other down? It's not just a desire to exercise one's muscles; if it were, they would do push-ups and sit-ups instead. It's not sadism or masochism. The pleasure is in the fighting, not the hurting or being hurt.

The solution to this puzzle is that play-fighting is a form of practice. Fighting is a useful skill, and practice will make you better—if you get in a lot of fights, you'll get better at it. But if you lose a fight, you might be crippled or killed, and even winners get broken fingers, smashed noses, and lots of pain. How do you get the benefits without suffering through the cost? The clever solution here is that animals who are kin or friends can use one another to improve their fighting skills while holding back so that nobody is hurt. This is why play-fighting has evolved.

In general, play is safe practice. You get better at something the more that you do it. But real-world experience can be costly, so people are drawn to involve themselves, in a safe way, in certain physical, social, and emotional situations. Sports are physical play; games are intellectual play; and stories and daydreams are social play, in which we vicariously and safely explore new situations.

Much of our play goes on inside our heads, and this helps us make sense of our hunger for aversive fictions. Just as play-fighting involves thrusting oneself into a situation that would be dangerous if real, our imaginative play often takes us into situations that include elements that would be unpleasant, sometimes terrible, if they existed in the real world. As the horror-writer Stephen King argues, we make up imaginary horrors to help us deal with real ones; this is "the tough mind's way of coping with terrible problems."

We are drawn, then, toward worst-case scenarios. The details of the scenarios are often irrelevant. It's not that we enjoy zombie films because we need to prepare for the zombie uprising. We don't have to plan for what to do if we accidentally kill our fathers or marry our mothers. But even these exotic cases serve as useful practice for bad times, exercising our psyches for when life goes to hell. From this perspective, it's not the zombies that make zombie films so compelling, it is that the theme of zombies is a clever way to frame stories about being attacked by strangers and betrayed by those we love. This is what attracts us; the brain eating is an optional extra.

Horror films are just one sort of practice. Some people avoid it, just as some never play-fight. But there are other ways to prepare for the worst, and we each pick our own poison. You might not like *Chainsaw Killers III*, but find yourself drawn to exploring the dimensions of loss by watching *Terms of Endearment* (mom dies of cancer) or *The Sweet Hereafter* (children, school bus, cliff).

Or you might stop to gawk at highway accidents. This vice was anticipated by Plato. In *The Republic*, he tells of Leontius who is walking in Athens and sees a pile of corpses, men who had just been executed. He wants to look at them, but turns away, struggles, at war with himself, and finally runs to the corpses and says to his eyes: "Look for yourselves, you evil wretches, take your fill of the beautiful sight!" The corpses are real enough, but they are safely seen at a distance, and the urge to look at them is the same urge driving us toward imaginary gore and imaginary death.

Paul Rozin has discussed other cases in which we willingly expose ourselves to controlled doses of pain. There is the uniquely human pleasure we get from spices such as chili and drinks such as black coffee. There is stepping into a too-hot bath, roasting in a sauna, courting nausea and fear on a roller coaster, or the self-infliction of mild physical pain, as with pressing your tongue against a sore tooth or putting a little bit of weight on a sprained ankle.

Can all of this "benign masochism" be explained by safe practice? Perhaps not—it is hard to see why we would we need to practice eating spicy foods or taking hot baths. These Rozin cases might have a more utilitarian explanation, something along the line of the awful old joke about the guy who was banging his head against the wall; when asked why he was doing it, he said, "It feels so good when I stop." For some of Rozin's examples, the initial pain might be worthwhile because it is outweighed by the later pleasure. We might grow to like the pain of stepping into a hot bath, because it is always followed by the bliss of when the temperature becomes just right.

NOT-SO-BENIGN MASOCHISM

We haven't yet considered those who engage in honest-to-God masochism, who have others beat and torture and humiliate them. These are unusual folk and a theory that works for those who enjoy *Friday the 13th* and five-alarm chili might not apply to them.

There is no shortage of possibilities. Some masochists may be jaded, so habituated to the humdrum that the adrenaline rush of pain and fear is required to capture their interest. Or, in cases of self-harm and self-mutilation, it might be a distress call, proof that they are sufficiently desperate to damage their body. Or maybe, as some have speculated, there is an odd form of learning here: pain leads to a blast of opiates that reduce the pain, but over time, some people might get more pleasure from the opiates than pain from the pain. This is the hot bath theory taken to an extreme.

Or perhaps it is self-punishment. A desire to punish is an early emerging and universal trait. In some recent research that I have done with the psychologists Karen Wynn and Kiley Hamlin, we find that children before their second birthday will punish (by taking away food) an individual who stole a ball from another. And there are many demonstrations, in the lab and in the real world, that adults will engage in what is called *altruistic punishment*. They will sacrifice something of their own, such as money, to punish an evildoer. Freud suggested that masochism is sadism directed inward; the idea here is similar—perhaps severe masochism is punishment directed toward oneself.

A fictional example that fits with this is the abused house elf Dobby in the Harry Potter series. He harms himself when he does something wrong: "Oh no no, sir, no . . . Dobby will have to punish himself most grievously for coming to see you, sir. Dobby will have to shut his ears in the oven door for this." It's not just fiction, though. In one clever study, undergraduates were invited to give themselves electric shocks by turning a dial. The interesting finding is that the intensity of the shock went up if, before being hooked up to the machine, they were asked to recall some sin, something they had done wrong in their lives.

One parallel between severe masochism and everyday masochism is that, in both, you need control over the intensity of the pain. The lover of spicy foods needs to have power over what's going into her mouth; the horror-movie fan gets to choose the movie and is free to close his eyes or turn his head. And in sadomasochism (S/M), it's critical for the person experiencing the M to have some sort of signal that means *Stop* and for the person doing the S to immediately respond. The signal is sometimes called, appropriately enough, a "safe" word.

The French philosopher Gilles Deleuze might be partially right, then, when he insisted that masochism isn't really about pain and humiliation, it's about suspense and fantasy. Control is essential, and this is what makes masochistic pleasure so different from ordinary pleasure. In a disturbing discussion, the writer Daniel Bergner describes how a horse buyer named Elvis chose to be basted with honey and ginger, tied to a metal pole, and roasted on a spit for three and a half hours. This is a lot of pain. My bet, though, is that if Elvis woke up one morning, stepped out of bed, and badly stubbed his toe, he wouldn't enjoy it at all, because it is not what he signed up for.

The ultimate test case here is going to the dentist. (What's the difference between a sadist and a dentist? Newer magazines.) One article on sadomasochism describes a woman with a high need for pain in S/M sessions with her boyfriend, but who hated going to the dentist. The boyfriend tried to get her to construe a dental exam as an erotic masochistic adventure, but failed. There was no getting around the fact that the dentist was necessary pain, not something she chose.

DAYDREAMING

The mind wanders. When our consciousness is not otherwise engaged, we mull over the past, schedule our vacations, humbly accept awards, win arguments, make love, and save the world. It is difficult to come up with a precise estimate of how much of our lives we spend at this, but in a set of studies performed about 30 years ago, people were beeped at a quasi-random schedule as they went about their day and were asked to record what they were doing when the beep went off. About half of their waking lives were spent doing some sort of daydreaming.

A recent fMRI study took this further, looking at brain activation in experimental subjects doing a repetitious task. The researchers found a network of regions in the brain that is active when people report that their minds are wandering and conclude that activation of this mind-wandering part of the brain is the default state. It only shuts down when people are doing something that demands their conscious attention.

Daydreaming involves the creation of imaginary worlds. You can imagine yourself in the woods or walking on a beach or flying. Here we are set designers. We are also casting directors and screenwriters, creating imaginary beings to populate these worlds, individuals who interact with us as if they were other people. An extreme version of this is manifested in schizophrenia, in which this other-self creation is involuntary and the victim of the disease believes that these selves are actual external agents such as demons, or aliens, or the CIA. But in the usual version, one controls these individuals and knows that they are self-generated, and every human who has the gift of speech sometimes exercises that gift by conversing with people who are not really there.

Sometimes a specific imaginary being sticks around, goes from a bit player to a regularly returning character. When it happens with children, we describe these alternative selves as imaginary friends or companions. The psychologist Marjorie Taylor has studied this phenomenon more than anyone else, and she points out that, contrary to some stereotypes, children who have such companions are not losers, loners, or borderline psychotics. If anything, they are more socially adept than children without these companions. And they are in no way deluded. Children are fully aware that the characters live only in their imaginations.

Long-term imaginary companions are unusual in adults, though they do exist—Taylor finds that authors who write books with long-standing characters often claim that these characters have wills of their own and get to have some say in their fates.

DAYDREAMS CAN give various sorts of pleasure. Our perfect control makes them the ideal venue for the sort of painful play we have been talking about. Many daydreams are masochistic. People imagine the worst: failure, humiliation, the death of loved ones. Then there is the simpler pleasure of simulating real-world delights. When daydreaming, we produce private movies in our heads in which we are the stars—with limitless budgets, a free hand in casting, great special effects, and no censors.

This raises a puzzle, though. If our daydreams are so good, why do we ever leave the house? Why do we seek out other imaginative pleasures and other real pleasures?

One weakness is that self-generated imagined experiences are less vivid than real ones. Imagine, as best you can, what it feels like to bite your tongue. Now bite it. See? Actual images on a screen can evoke sexual arousal, horror, or disgust with an intensity that self-generated imagery cannot match.

The second weakness is that in the movies of my imagination, I am director and screenwriter. This is bad news, because I am not a gifted director or screenwriter. Steven Spielberg and Pedro Almodóvar can direct better movies than I can; the Coen brothers are better screenwriters. Shakespeare can write better plays. They are able to think up some pleasurable fantasy for me that I'm not creative enough to think up myself. Or some fascinating interplay between people. Or even some suitably painful masochistic experience.

The third weakness of daydreams is their lack of limit. As the psychiatrist George Ainslie put it, daydreams suffer from a "shortage of scarcity." This diminishes the masochistic power of daydreams, because one can never surprise oneself in an unpleasant way. It also diminishes the joy of simulating a real-world pleasure, because so many real-world pleasures involve some loss of control and in a daydream you have perfect control. There is nothing you can't do; all of your failures happen because you chose to fail, and so where's the value in winning?

The point was appreciated in a classic *Twilight Zone* episode in which a violent thug dies and finds himself in a place in which his every wish is satisfied. He is shocked to be in paradise and at first has a wonderful time. But he gets frustrated and bored, and after a month, tells his guide: "I don't belong in heaven, see? I want to go to the other place." The guide responds: "Whatever gave you the idea that you were in heaven, Mr. Valentine? This *is* the other place!" Cue maniacal laughter.

Daydreams are the opposite of dreams, then, because in dreams you typically have no control at all. This means that a good dream can be more pleasurable than a good daydream, while a nightmare can be terrible indeed.

There are clever fixes to improve daydreams. The philosopher Jon Elster points out that one can daydream with a friend. Part of the bonus here is that the friend might think of clever new scenarios, but the real benefit is that another person constrains the situation. One has to deal with another's competing interests and desires, a form of constraint that can ramp up the pleasure.

Then there is immersing oneself in virtual worlds, from the stripped-down physical worlds of a racing or flight simulator to the full-blown social universes of Second Life or World of Warcraft.

This can be seen as an enhanced form of daydreaming: you are an agent in an unreal world, but this world is constrained, and you cannot always get what you want. You can also benefit from the imaginary resources of others—there are experiences available in Second Life, for instance, that I would have never thought to provide for myself.

Such worlds are increasingly popular, larger than many countries. There are people who spend most of their waking hours within them, and I suspect that this is going to become more common as the technology improves. A psychologist I know asked one of her research assistants to try out one of these worlds and report on what it is like and how people behave there. The research assistant never came back; she preferred the virtual life to the real one.

Imagination changes everything. It evolved for planning the future and reasoning about other minds, but now that we have it, it is a main source of pleasure. We can partake in experiences that are better than real ones. We can delight in the minds that create imaginary worlds. And we can use the pain potential of the imagination to play at unpleasant realities, mentally practicing with scenarios that are both safe and terrible.

There will be more of this to come. Virtual worlds will expand, making interactive daydreaming more attractive, and technological improvements will blur the distinction between reality and imagination. One day we will have holodecks and orgasmatrons—or at least more advanced television sets.

Imagination has its limits, though. Our ambitions go beyond the acquisition of experiences; they extend outside the head. Someone who trains for a marathon doesn't merely want the experience of running a marathon or the belief that she has run a marathon, she wants to run a marathon. All else being equal, flying a plane is better than a flight simulator; real sex is better than masturbation; real gossip is better than the clever imagined dialogues of characters on television. The pleasures of the imagination are a core part of life—but they are not enough.

8

WHY PLEASURE MATTERS

FOR MOST OF THE HISTORY OF OUR SPECIES, THERE WAS NO television or Internet or books. Our ancestral environment had no McDonald's, birth control pills, Viagra, plastic surgery, nuclear weapons, alarm clocks, fluorescent lighting, paternity tests, or written codes of law. There were not billions of people.

Our minds are not modern, and many of our woes have to do with this mismatch between our Stone Age psychologies and the world in which we now live. Obesity is a simple example of this. Food was hard to get for most people for most of human history. Even a few hundred years ago, the average European family spent over half of its budget on food, and it wouldn't get much for the money—the daily caloric intake of an eighteenth-century Frenchman was equivalent to that of a contemporary citizen of a malnourished African nation. In a world in which food is scarce, it is smart for an animal to eat when it can and store up the fat, and suicidal to pass up the chance for sweet fruit or fresh meat. But many humans now live in environments in which food is cheap, plentiful, and cleverly manufactured to be maximally flavorful. It is difficult—impossible for many of us—to resist the Darwinian imperative to gobble it all up.

As another example, it would be smart to treat the insults and provocations of strangers—rude behavior on the highway, nasty remarks on the Internet—as irrelevant. There's no payoff to getting mad. But our minds are not evolved to think about strangers, and we obsess, needlessly, about what people think of us and how these insults will diminish us in the eyes of others. This is why we have road rage and blog wars.

Finally, we have evolved in a world of lions and tigers and bears; of plants and birds and rocks and things. We get pleasure and fulfillment from the natural world. Many modern humans miss out on this, as we spend our days in constructed environments. The biologist E. O. Wilson has argued that this estrangement from nature is bad for the soul: "[We] descend farther from heaven's air if we forget how much the natural world means to us." Several studies now show that even a limited dose of the natural, such as a chance to look at the outside world through a window, is good for one's health. Hospitalized patients heal quicker; prisoners get sick less often; spending time with a pet enhances the lives of everyone from autistic children to Alzheimer's patients.

These mismatches are interesting and important, and they are the focus of much research and theorizing in evolutionary psychology. What scholars sometimes miss, though, is that we are not innocent bystanders. We are not like rats dropped into a psychologist's maze or elephants thrust into a circus. We made this unnatural world. We invented the Big Mac and the Twinkie, the freeway and the Internet and the skyscraper, government and religion and law.

This book so far has been about what we like and why we like it. In this brief final chapter, I turn to some implications of the essentialist nature of pleasure, and discuss its influence on the world that we now live in.

ESSENTIAL NONSENSE

Arthur Koestler tells the story about a 12-year-old girl, a daughter of a friend, who is taken to Greenwich Museum and later asked to name the most beautiful thing there. She says it is Admiral Lord Nelson's shirt. As she puts it, "The shirt with blood on it was jolly nice. Fancy real blood on a real shirt which belonged to someone really historic."

You can almost hear Koestler sigh as he writes: "We can no more escape the pull of magic inside us than the pull of gravity." Magic is a loaded word here, with its implication of irrationality, but maybe this is fair. It is one thing to prefer a certain chair because it is more comfortable or to enjoy a painting because one is struck by its beauty. This makes sense. But isn't it weird that we enjoy things—like a dead man's shirt—not because of anything that they can do for us, and not because of any tangible properties that they have, but rather because of their histories, including the invisible essences that they contain? Essences that don't really exist! In fact, isn't this book so far a chronicle of human silliness—silliness about food, about sex, and so on? Isn't this a long argument that pleasure is affected by factors that shouldn't really matter?

Some psychologists would say so. My collaborator Bruce Hood

makes a point similar to Koestler's, arguing that these perverse attachments should be lumped in with worries about black cats and haunted houses. They are unreasonable. In his discussion of originals and forgeries, he writes: "When art critics and gallery owners talk about the essence of a piece of art, they are talking essential nonsense." And in an experimental article that explores the liking of everyday objects, Hood and his colleagues contrast "rational economic decisions," which have to do with real-world utility, with "clearly irrational judgments"—by which they mean valuing an object for sentimental reasons, such as a child's attachment to a security blanket.

Demonstrations of human irrationality are nothing new. Consider the research done by Amos Tversky and Daniel Kahneman, for which Kahneman won the Nobel Prize in Economics in 2002. Tversky and Kahneman find that we are often quite poor at logical deduction and probabilistic reasoning. We might pay $99.99 for the new set of speakers but walk away if the price is $100.00; we obsess over the dangers of guns in the house but are indifferent to the (far more serious) threat of swimming pools. Our imperfection is not surprising. We are animals, not angels. Our minds have been shaped by natural selection to reason in useful ways about the world, but evolution is a satisficer, not an optimizer. Recall also that our minds have evolved for a different world than the one in which we now live. It makes sense, then, that we might now reason in ways that are not just imperfect, but ineffective. As the psychologist Gary Marcus has so nicely argued, our brain contains "kluges."

Is essentialism one of them? Certainly people have mistaken essentialist beliefs. Sex with a virgin won't cure AIDS and eating the corpse of a person who speaks English won't improve your own English. Human groups, such as blacks and Jews, don't have essences that clearly distinguish them from other groups. In a recent court case in Britain, Procter & Gamble argued that Pringles are not potato chips (and hence not subject to value-added tax rules) because they do not contain enough potato to have the "essence of potato." The Supreme Court of Judicature rightly rejected this claim, pointing out that this sort of Aristotelian notion doesn't apply here—Pringles don't have essences.

Much of what we think about essences is mistaken, then. But this doesn't mean that the general essentialist intuition is mistaken. As we discussed at the start of this book, there really is a deeper reality to things: Tigers aren't merely a sort of animal with a certain appearance; tigers have deeper properties that make them tigers, having to do with DNA and evolutionary history. Gold isn't just a substance of a certain color; what makes gold gold has to do with molecular structure. Individuals have essences too. Two newborns might be hard to tell apart, but if one of them is your child and the other isn't, that's an invisible genetic fact that really matters. Essences exist, and it makes sense for us to be attuned to them.

But why should the girl care whose blood was on Nelson's shirt? What is going on with a person who paid about $50,000 for a tape measure from the Kennedy household? The buyer was Juan Molyneux, a Manhattan interior designer, and he remarked: "When I bought the tape measure, the first thing I measured was my sanity."

I think he is being too hard on himself. He would be mistaken if he were confused about the tape measure—if he thought it had magical powers, for instance. But if he simply likes it because of who owned it, this is just a matter of taste. It is neither rational nor irrational. If you like vanilla and I like chocolate, we have a disagreement but neither of us is being unreasonable. Similarly, if Sarah likes tape measures based mostly on their sensory appeal and everyday utility and Juan likes tape measures based mostly on their history, it is confused to say that Sarah is smarter (or more virtuous, or more reasonable) than Juan—or vice versa.

This point applies more generally. In his discussion of individuals with unusual sexual desires, Daniel Bergner interviewed a masochist, a man in his midforties who left Wall Street to spend more time with his children. The interview took place in the workshop of a dominatrix, with the man strapped on his back onto a worktable, wearing a latex bodysuit and a mask with an opening only for his mouth. Attached to the shaft of his penis was a conductive ring that led to a small machine. The dominatrix wired up the machine to generate electrical shocks at the sound of speech, making Bergner complicit in the man's torture. But when Bergner gently asked about his childhood experiences, the masochist denied that he was unusual: "I was never raped by homosexual dwarves. Is this a weird way to deal with life? Consider the man who bought Mark McGwire's seventieth home-run ball for three million dollars. Who's weirder?"

I think both men are pretty weird, actually. Still, neither is *mistaken* in any sense. One can imagine a species much like us except that their brains are wired up in a different way, so that they are not natural-born essentialists and hence are indifferent to the deeper nature of things. Such creatures would not experience many of our pleasures. They would happily exchange their prized wedding rings for duplicates. They would not collect autographs or mementos, and the young of that species would not get attached to security objects such as soft blankets. They would not get the same pleasures from art and fiction and even masochism, because they wouldn't care about the human act of creation underlying these experiences. Such individuals would not be smarter, or dumber, or more or less rational than us—they would just be different.

THERE IS plenty of room to be judgmental. The issue isn't true or false, rational or irrational. It is right and wrong. Some pleasures are immoral ones. Some of them lead to human suffering. Even if it is not unreasonable to take pleasure in sex with young children, say, it is an immoral pleasure, one to be discouraged. If your love of food leads you to destroy your own body or take what belongs to others, you are a glutton, and this is to be discouraged as well.

Some of our essentialism leads us to behave in ways that are immoral. We have discussed such cases earlier, including gruesome examples such as the murder of children for their flesh and the ugly obsession with female virginity.

Essentialism can also drive us to become obsessed with material objects and to ignore the needs of real people. Economists such as Robert Frank and Richard Layard and evolutionary psychologists such as Geoffrey Miller have argued that the obsession that many of us have with acquiring luxury goods has a social cost, and that society would do better if such acquisitions were blocked or discouraged. The philosopher Peter Singer has made this argument in sharper terms, outlining the moral issues that arise when we spend our money on expensive clothing and cars instead of using it to save the lives of starving children. If we were a nonessentialist species, we would value certain material things a lot less and perhaps value actual people a little bit more. There is a cost to our pleasure.

SEEKING OUT ESSENCES

People often insist that, in some superficial and noninferential way, the expensive bottled water is just plain tastier than the stuff from the tap or that the original Chagall just looks superior to the fake—any discerning person should be able to tell. In such cases, we are unaware of the depth of our pleasures.

In other domains, though, we are explicitly aware of our interest in essences. This interest shows up in the curiosity that many people have about the underlying intentions of an artist or storyteller and, in particular, about whether a story is real or make-believe. It shows up when we are engaged in love and romance. Some of us are very interested in how old someone really is, as opposed to how old he or she looks, and there is an intense curiosity about who is using plastic surgery and Botox and hair plugs and the like. In general, modern attempts to obscure what we see as the real person often trouble us, morally and aesthetically, and this is reflected in the unease we feel about various physical and psychological enhancements.

Or consider the appeal of nature. We pay to live close to oceans, mountains, and trees—a Manhattan apartment with a view of the greenery of Central Park is worth far more than one facing the other way. Office buildings have atriums and plants; we give flowers to the sick and the beloved, and return home to watch Animal Planet and the Discovery Channel. We keep pets, which are a weird combination of constructed things (cats and dogs were bred for human companionship), surrogate people, and conduits to the natural world. And many of us seek to escape our manufactured environments whenever we can—to hike, camp, canoe, or hunt.

When it comes to nature, we want the real thing; we are uncomfortable with substitutes. There is a fortune to be made, for instance, by building a robot that children would respond to as if it were an animal. There have been many attempts, but they don't evoke anywhere near the same responses as puppies, kittens, or even hamsters. They are toys, not companions. Consider also a study by the psychologist Peter H. Kahn, Jr., and his colleagues. They put 50-inch HDTVs in the windowless offices of faculty and staff to provide a live view of a natural scene. People liked this, but when they were tested using physiological measures of heart-rate recovery from stress, watching the HDTVs was shown to be worthless, no better than staring at a blank wall. What did help with stress was giving people an office with an actual plate glass window overlooking actual greenery. I think we are searching for actual nature, and our understanding of how important this is to us underlies some of the anxiety that we feel about nature's loss.

These are examples of domains in which at least some of us are aware of our essentialism. But there is something more. Many people, perhaps all of us, are consciously aware that there is something more to the world than what we perceive. There is an underlying reality that we want to make contact with.

This is one motivation behind the enterprise of science. Several years ago, the biologist Richard Dawkins wrote a book called *Unweaving the Rainbow*. The title was a reaction to Keats's worry that Newton had destroyed the poetry of the rainbow through his physics. Dawkins argues that this isn't so: "The feeling of awed wonder that science can give us is one of the highest experiences of which the human psyche is capable. It is a deep aesthetic passion to rank with the finest that music and poetry can deliver. It is truly one of the things that makes life worth living." Dawkins is talking here about the pleasures of science, the joy of this way of making contact with the deeper nature of things.

Now, science as an institution hasn't been around that long, and some societies still don't have it. Even in the West, there are probably more foot fetishists than scientists. But Dawkins's argument is meant to extend more broadly to those who are consumers or potential consumers of scientific insights, and I think the public appetite for books of the sort that Dawkins writes is proof that many people who are not themselves practicing scientists are interested in, and get some pleasure from, knowing about the deeper nature of things.

Still, science is not the most popular way to reach out to a transcendent reality. Most people scratch this itch in a different manner. They achieve the pleasure of "awed wonder" without having to mull over the details of Mendelian genetics or the periodic table or the wave-particle duality of electrons. Rather, the desire to contact the transcendent is satisfied though a different social enterprise—religion.

People mean different things when they talk about religion. A popular approach is to see religions as belief systems, characterized by certain claims about how things really are. This was the approach of the anthropologist Edward Burnett Tylor, who, in 1871, wrote that the "minimum definition of religion" is a belief in the existence of spiritual beings—in gods, angels, spirits, and the like. If you have such beliefs, you are religious. I think this is a sensible way of capturing what religions all have in common, and my last book was in part an exploration of where such beliefs come from. One might also think about religions as sets of practices and specific group affiliations. To be a Christian, for instance, is to engage in certain rituals and to affiliate with certain people.

From this standpoint, the interesting psychological questions concern the nature of the rituals and why people choose to participate in them, as well as the ways in which people form distinct social groups.

I think, though, that there is more to religion than belief and ritual and society. There is something more basic that all religions share and which spills over as well to what is often described as *spirituality*. This is the notion that there is more to the world than what strikes our senses. There is a deeper reality that has personal and moral significance. The sociologist and theologian Peter Berger talks about the central assumption "that there is *an other reality*, and one of ultimate significance for man, which transcends the reality within which our everyday experience unfolds." William James, in *Varieties of Religious Experience*, writes that religion "consists of the belief that there is an unseen order, and that our supreme good lies in harmoniously adjusting ourselves thereto." This is what scholars are talking about when they discuss the sacred versus the profane: the profane is the everyday world; the sacred is the other reality that people hunger for.

The underling reality of religion is different from that of science in a couple of ways. For one thing, science tells us, as the physicist Steven Weinberg once put it, that the universe is pointless. It has no interest in our success or happiness; it provides no moral guidance. In contrast, the deeper reality expressed in religion is full of meaning or morality and love. For another, while science can tell us about the deeper reality (through tools like microscopes) and can sometimes even manipulate it (through techniques such as gene splicing), religion has even more powerful mojo, because it provides tools that work at an experiential level.

This is one of the functions of ritual. For some rituals, the deeper reality—or the supernatural—can somehow become manifest in the objects of the ritual, as in the Eucharist in which the wine and the wafer become the blood and body of Christ. (Scholars, such as Walter Benjamin and Ellen Dissanayake have argued that this is similar to what goes on in the creation of art.) For other rituals, a person can become directly linked to this deeper reality, as in prayer or meditation or some other sort of transcendent personal experience. Such experiences can have immense significance.

Religion and science are social institutions that exist in part to satisfy the interest that people have in the transcendent, but the interest itself predates these institutions. You don't need religion to have ritual, for instance; children can create them spontaneously. Some of this might be reflexive association—I won the ball game when I wore my lucky socks, now I wear them to every game—but some of it may reflect a deeper belief system. In their extensive study of the belief systems of thousands of children, the folklorists Peter and Iona Opie conclude that the need to create such rituals is part of human nature and that children "have an innate awareness that there is more to the ordering of fate than appears on the surface."

Similarly, one doesn't have to be a scientist to be interested in what things are made of and where things come from. The psychologist Alison Gopnik makes a good case that the terrible twos can be explained as the insanely curious toddler doing "experiments" on the world, acting on people and objects and attending to the results. And there has long been a broader movement in developmental psychology, emerging from the work of the psychologist Susan Carey, in which the cognitive development of children is seen as analogous to scientific progress.

A critic might wonder, though, how much of these prereligious and prescientific impulses reflect essentialism in a strong sense, as opposed to a more general desire of children to better manipulate and understand the world. I wonder about this myself. I am convinced by the experimental research summarized in the first chapter that even preschool children are commonsense essentialists, in that they tacitly believe that categories and individuals have hidden and invisible essences. But do they really have a specific desire to contact these essences? Does their essentialism give them *pleasure*? I think it's too early to tell.

The evidence is clearer with adults, though. Even those who explicitly reject religious belief show signs of the transcendent impulse. They are not blind to the attraction of a deeper reality; they just resonate to this attraction outside of the bounds of organized religion. As an illustration, consider the view of some prominent modern-day atheists. I have already discussed how Richard Dawkins wrote a book about the transcendent appeal of scientific inquiry. Sam Harris is well known for his attack on the monotheistic faiths, but he is strongly enthusiastic about Buddhism, describing it as "the most complete methodology we have for discovering the intrinsic freedom of consciousness, unencumbered by any dogma." And Christopher Hitchens, author of *God Is Not Great*, has spoken about the importance of the "numinous"—which usually refers to the experience of contact with the divine—and has argued that one can experience it without religious or supernatural belief. He suggests that humans rely on the numinous and the transcendent, and says that he personally wouldn't trust anyone who lacked such feelings.

Even hard-core rationalists share this hunger for the transcendent, then. If you're looking for individuals who are blind to this, you might be looking at the wrong species.

AWE

This experience of the transcendent may be connected to the fascinating and little-understood emotion of awe.

There are many triggers to awe. The psychologist Dacher Keltner notes that the classic cases involve encountering the divine. Paul's conversion on the road to Damascus, in which he was blinded by the light, is a famous example. A more detailed account is at the end of the Hindu Bhagavad Gita, in which the hero Arjuna asks Krishna if he can see the universe for himself, and so Krishna gives him a "cosmic eye." Arjuna then sees gods and suns and infinite space: "Things never before have I seen, and ecstatic is my joy; yet fear-and-trembling perturb my mind." This is awe.

Over time, scholars began to see this emotion as connected to other, nondivine experiences. In 1757, Edmund Burke talked about the sublime—an awelike reaction that we can feel from hearing thunder, viewing art, and listening to a symphony. For him the two ingredients of the sublime are power and obscurity. In our times, its scope is bigger still. When Keltner asks undergraduates at the University of California at Berkeley to tell him about their experience of awe, they talk about music, art, powerful and famous people, sacred experiences, certain perceptual experiences, meditation and prayer. They tell him about how they felt when the Red Sox won the World Series, or during their last experience of sex, or when they were lifted aloft in a mosh pit, or when they were high on LSD.

What do these experiences have in common? Keltner, working together with the psychologist Jonathan Haidt, emphasizes the features of vastness—physical, social, intellectual, and otherwise—and of accommodation, in which we struggle to deal with this vastness. He notes that when we feel awe, we feel small, and this corresponds to certain physical responses that sometimes accompany the experience, such as bowing, kneeling, or curling into a ball. (When Paul saw the light on the way to Damascus, he fell to the ground.)

Awe is a mystery from an evolutionary perspective. Keltner suggests that at its core, awe is a social emotion; it corresponds to a "sense of reverence for the collective." Its primary trigger is powerful people who unite the community, and we diminish ourselves and are subservient to these awe-inspiring others. In this regard, awe is similar to social emotions such as loyalty to the in-group and fear and hatred of the out-group. It is a social adaptation.

This is an interesting hypothesis, but it has some gaps. For one thing, it is not clear why awe is elicited by entities and experiences that have nothing to do with holding together a collective, such as the Grand Canyon or impressionist artwork or getting stoned on acid. For another, there is something fishy about the claim that we have evolved an emotion specially geared to being blown away by the powerful. Such people are not saints. What they want from us is not necessarily subservience for the good of the community—it's subservience to *them*. They want our mates, our children, and our resources. Why then would we be wired up to cheerfully hand it all over? How could such a response evolve? If you think about two hominids, one inclined to fall to its knees and give it all up for the great leader, the other more cynical, it's not clear why the hero-worshipper's genes would be more likely to prosper.

Keltner would see this as too cynical. He is a fan of awe, seeing it as an emotion "that transforms people, energizes them in the pursuit of the meaningful life and in the service of the greater good." I think the world would be better off if awe didn't exist. We would be better off if we would cold-bloodedly assess the abilities and goals of prospective leaders and weren't so prone to swoon. When Keltner thinks of people who have been objects of awe, he thinks of deserving sorts such as Gandhi and the Dalai Lama. I think of Hitler and Stalin, along with countless other tin-pot dictators, polygamous religious zealots, and Machiavellian creeps, all willing to exploit this psychological blind spot.

If awe isn't a social adaptation, what's the alternative? One tentative hypothesis—consistent with the work of Keltner and Haidt—is that it is not an adaptation at all, but an accident. People are drawn to seek out the deeper essence of things; we are curious, and the payoff for learning more is a click of satisfaction. Indeed, in an interesting paper called "Explanation as Orgasm," Alison Gopnik makes the connection between the satisfaction of orgasm as a spur to more sex and the satisfaction of a good explanation as a spur to further exploration. But you can have too much of a good thing. Perhaps the feeling of awe is what we get when the system is overwhelmed; there is too much to process, too much physical vastness, or seemingly divine power, or human virtuosity.

IMAGINE

The capacity to think about worlds that don't exist is a useful human power. It allows for the contemplative assessment of alternative futures, something indispensable for planning our actions; it lets us see the world as others see it (even if we know that they are wrong), which is essential for human acts such as teaching, lying, and seduction. And, combined with our essentialism, it leads to pleasures that are central to our modern lives.

For one thing, it makes fiction and art possible. It is obvious that the creator of a story or artwork needs imaginative powers, but it is true also for the audience. The pleasure of fiction isn't accessible at all unless you have the imaginative powers to create an alternative reality. And the pleasure of artwork often involves an interpretive leap, an educated guess about what went on during the creative process. Aesthetic pleasure is to some extent an act of reverse engineering, except that instead of physically taking apart the object to see how it was made, you do it in your mind. Without the capacity for imagination, you might enjoy an attractive splash of colors on a canvas, but you would never enjoy art in the same way that normal people do.

Imagination also makes science and religion possible, because they both explore realities that are not present to the senses. Nothing would be left of these human practices if we couldn't imagine a hell below us and a heaven above, or if we couldn't think about a perfect sphere or infinite space. We'd be lost without the capacity to appreciate that a liquid looks like wine but is really the blood of Christ or that a rock is really composed of tiny particles and fields of energy. Indeed, what we are doing right now—thinking about what we would lose if we had no powers of the imagination—is itself an exercise of the imagination.

In science, one specific role of the imagination is in the aid of what philosophers have dubbed "thought experiments," in which one illustrates or tests a scientific hypothesis by imagining a certain situation. Galileo used a thought experiment involving dropping pairs of stones off a tower to refute Aristotle's claim that heavier objects fall faster; Einstein used one involving a moving train to illustrate the theory of relativity.

For religion, there is special emphasis on stories; religious texts are full of them. Stories make the religious ideas stick over time—they are far more memorable than lists of facts. They make the ideas appealing to children, given the pleasure that children take in fiction and make-believe.

Stories may play another role in religion. There is a play-acting element to much of religion, where you pretend that something is true. Now, it would be wrong, and offensive, to claim that when the devout say that they are consuming the blood and body of Christ, they are playing, like a four-year-old shooting down criminals with his finger or imagining that a banana is a telephone. Often religious claims are sincere beliefs about reality, akin to a scientist's belief that water is made out of molecules; you can't see it, but it's true.

But religions make numerous claims, and they shouldn't all be taken equally seriously. To shift to a ritual from my own tradition, in the Passover ceremony we open the door so that Elijah can enter and drink a cup of wine from the table. This is pure play, a children's story, and afterward the cup is tossed out or poured back into the bottle. Perhaps there are some Catholics who think of the Eucharist in a similar regard, as a ritual without any metaphysical implication. Or consider the act of praying. For some, this is actual communication with a divine being; for others, it is little more than a nervous tic. And for many, it is somewhere in between.

It is the in-between cases that are particularly interesting. The situation here is reminiscent of what the psychoanalyst Donald Winnicott said about babies' relations to transitional objects like teddy bears and soft blankets. He claimed—plausibly, see Chapter 4—that these were substitutes for the mother, or perhaps just for her breast. But what do babies themselves think of them? Do they recognize that they are substitutes, or do they think that they are actually mothers/breasts? Winnicott has an odd remark about this: "Of the transitional object, it can be said that it is a matter of agreement between us and the baby that we will never ask the question: 'Did you conceive of this or was it presented to you from without?' The important point is that no decision on this point is expected. The question is not to be formulated."

In other words: don't ask. I think Winnicott's remark captures the ambiguity that many people feel with regard to their religious beliefs. They have an odd and fragile status. For science too, there are questions that arise about certain more theoretical constructs. Are quarks and superstrings real or convenient abstractions? Some would advise: don't ask.

In any case, imagination and transcendence are intimately related. Imagination serves as a tool through which to achieve certain forms of transcendent pleasure. We have the power not only to try to connect to a deeper reality, but to envision what this reality might be.

This power exists in children as well. My favorite story along these lines was told by the educator Ken Robinson about a classroom interaction that he heard about. There was a six-year-old girl sitting with her arms curled around a piece of paper, intensely absorbed in her drawing. Her teacher waited for more than 20 minutes and then went up to the girl and asked what she was drawing. Without looking up, the girl said, "I'm drawing a picture of God."

The teacher was surprised and said, "But nobody knows what God looks like."

And the girl said, "They will in a minute."

NOTES

PREFACE

xiii Spandrels: Gould and Lewontin 1979.

xiv The narrowness of psychology: Rozin 2006.

1. THE ESSENCE OF PLEASURE

1 Van Meegeren's story: Dolnick 2008, Wynne 2006.

5 Romanes on pleasure and pain: Quoted by Duncan 2006.

5 Happy people: Pinker 1997, p. 387.

5 Risen apes: Quoted by Jacobs 2004.

7 Up for grabs: Menand 2002, p. 98.

8 Seymour's story: Salinger 1959, pp. 4–5.

9 Essentialism: For philosophical foundations, see Kripke 1980; Putnam 1973, 1975; for psychological foundations, see Bloom 2004, Gelman 2003, Medin and Ortony 1989.

9 Locke on essence: Locke 1690/1947, p. 26.

10 Still a tiger: Keil 1989.

10 Artifact essentialism: Bloom 1996, 2000, 2004; Medin 1989; Putnam 1975.

10 Language and essentialism: Bloom 2000.

11 Borges list: Cited by Ackerman 2001, pp. 20–21.

11 Basis of natural order: Gould 1989, p. 98.

11 The weight of nouns: Markman 1989.

11 Not a racist: Gelman 2003.

12 Essentialism is true: Bloom 2004; see also Pinker 1997.

12 Minimal groups: Tajfel 1970, 1982.

13 Mitochondrial Jew: Gelman 2003, p. 89.

13 Half Irish: Gil-White 2001.

14 Boys have penises: Gelman 2003, p. 3.

15 Homer was not essentialist: Fodor 1988, p. 155.

15 Baby generalization: Baldwin, Markman, and Melartin 1993.

15 Appearance vs. essence in children: Gelman and Markman 1986, 1987.

16 Appearance vs. essence in even younger children: Gelman and Coley 1990; Graham, Kilbreath, and Welder 2004; Jaswal and Markman 2002; Welder and Graham 2001.

16 Dog insides: Gelman and Wellman 1991.

16 Common name for same internal stuff: Diesendruck, Gelman, and Lebowitz 1998.

16 Transformation: Keil 1989.

17 A hurter: Gelman 2003.

17 Carrot-eater study: Gelman and Heyman 2002.

17 Essentialism as special to biology: Atran 1998.

17 Essentialist about artifacts: Diesendruck, Markson, and Bloom 2003; see Bloom 2004 for review and discussion.

17 Boy instinct: Gelman and Taylor 2000.

17 From biology to socialization: Smith and Russell 1984; see also Hirschfeld 1996.

18 Penny example: Dennett 1996.

18 Children as vitalists: Inagaki and Hatano 2002.

19 Axe: E-mail from Emma Cohen, June 11, 2009.

19 Sterilization and value: Newman, Diesendruck, and Bloom under review.

20 No eye contact: http://www.happiness-project.com (search for "darshan").

20 Reborn a tampon: Hood 2009.

20 A noble form of cannibalism: Kass 1992, p. 73.

20 Property transfer: Sylvia and Nowak 1977.

21 Searching for the Dalai Lama: Bloom and Gelman 2008.

21 The story of the testing procedure: From Gould 1941 and Wangdu 1941; quotations are from Gould, p. 67, and Wangdu, p. 18.

23 Hard to question psychological phenomena: Cosmides and Tooby 1994.

23 Questions only a psychologist would ask: From James 1892/1905, p. 394.

2. FOODIES

25 Story of Meiwes and Brandes: L. Harding, "Victim of cannibal agreed to be eaten," *The Guardian*, December 4, 2003.

26 Cannibalism keeps a person close: Smith 1995.

27 Dahmer's anxious question: Smith 1995.

28 Chili on the breast: Rozin and Schiller 1980.

28 Humans as omnivores: Rozin 1976.

28 The tastes of supertasters: Bartoshuk, Duffy, and Miller 1994.

29 Explaining food preferences: Rozin and Vollmecke 1986.

30 Optimal foraging theory: Harris 1985.

31 Sudden food aversions: Rozin 1986.

31 Little relationship between parents and children on food preference: Birch 1999, Rozin and Vollmecke 1986.

32 Social learning from peers: Harris 1998.

32 Babies' learning about food from similar people: Shutts et al. 2009.

32 People taste like Spam: Theroux 1992.

33 Why insects are loathsome: Harris 1985, p. 154.

33 Disgust as an evolved response to rotting meat: Pinker 1997.

33 Retching and vomiting through the imagination: Darwin 1872/1913, p. 260.

34 The development of disgust: Bloom 2004; see also Rozin and Fallon 1987.

34 Giving children dog feces: See Rozin, Haidt, and McCauley 2000 for review.

34 Not fit for drinking: Siegal and Share 1990.

34 Miller's fastidious children: Miller 1997.

35 Getting adults to eat disgusting things: Smith 1961.

36 Cannibalism as a myth: Arens 1979.

36 Meat as desirable food: Rozin 2004.

36 Babies are delectable: Hrdy 2009, p. 234.

36 Two ways to be a cannibal: Lindenbaum 2004.

37 Snorting Dad: "Keith Richards says he snorted his father's ashes," April 4, 2007, http://www.msnbc.msn.com/id/1793369.

38 Cannibal dialogue: Harris 1985, p. 206.

39 You are what you eat: Nemeroff and Rozin 1989.

39 Blood libel against Catholics: Rawson 1985.

39 Eats my flesh; drinks my blood: John 6:54 (King James Version).

39 Loving, hungry monsters: Sendak 1988.

40 Placenta recipe: To find online, search for "Want a slice of placenta with that?"

40 Placenta as a TV dinner: Hood 2009.

40 Muti killing: Taylor 2004.

41 Killing of albinos: J. Gettleman, "Albinos, long shunned, face threat in Tanzania," *New York Times*, June 8, 2008.

41 Gandhi on the soul of animals: Coetzee 1995.

41 Cures for impotence: McLaren 2007.

41 Meat and manliness: Rozin 2004.

42 $15 billion a year on bottled water: Fishman 2007.

42 Natural over artificial: Rozin 2005.

42 The problem with natural foods: Pollan 2006, pp. 96–97.

43 Costly signals as an explanation for certain tastes: Cowen 2007, Frank 2000, Miller 2009.

44 Nevins's attempt to find the Perrier: Fishman 2007.

45 What you think affects what you taste: See Lee, Frederick, and Ariely 2006 for review.

45 Coke, Pepsi, and the brain: McClure et al. 2004.

45 Wine studies: See Lehrer 2009 for review.

46 Enjoying dog food: Described in Bohannon 2009.

47 Beer study: Lee, Frederick, and Ariely 2006.

48 Wine on the brain: Plassmann et al. 2008.

48 Cheese or body odor?: de Araujo et al. 2005.

51 Masochist recipe: Michaels 2007.

51 Benign masochism: Rozin and Vollmecke 1986.

52 Eating and humanity: Kass 1994.

52 Eating in the car: Pollan 2006.

53 Morality taking the place of etiquette: Appiah 2008, Pinker 2008. Appiah quotation is from Appiah 2008, pp. 245–46.

3. BEDTRICKS

55 Bedtricks: Doniger 2000.

56 Fear of a bedtrick: Example from McEwan 2005.

57 Laban's bedtrick: Genesis 29:25 (King James Version).

57 Jewish wedding ritual: Thanks to Murray Reiser for pointing this out to me.

58 Sex strategies of the male toad: Dekkers 2000, cited by Doniger 2000; Doniger quotation is from Doniger 2000, p. 130.

60 Parental investment: Trivers 1972, Clutton-Brock 1991; see Diamond 1998 for an accessible review.

62 Human sexuality: Diamond 1998.

62 Sex differences in sexual psychology: See Pinker 2002 for review.

64 Hidden ovulation: But see Miller, Tybur, and Jordon 2007 for evidence that men have *some* sensitivity to the timing of female ovulation.

64 Theories of the evolution of concealed ovulation: Diamond 1998.

65 Faces light up the brain: Aharon et al. 2001.

65 Babies like pretty faces: Langlois, Roggman, and Reiser-Danner 1990; Slater et al. 1998.

65 Beauty is arbitrary: Darwin 1874/1909.

65 What makes a face attractive: See Rhodes 2006 for review.

65 Average faces look good, even for babies: Langlois and Roggman 1990; Langlois, Roggman, and Reiser-Danner 1990.

66 Some attractive faces aren't average: Perrett, May, and Yoshikawa 1994.

66 Looks matter more to men: Buss 1989.

66 Men and women agree on attractiveness: Langlois et al. 2000.

66 Ovulating females like hypermale faces: Johnston et al. 2001, Jones et al. 2008, Penton-Voak et al. 1999.

67 An arousing head on a stick: Boese 2007.

67 Foot fetishists: Bergner 2009.

69 Sexy, smelly T-shirts: Wedekind and Füri 1997.

69 The attractiveness of women seen in class: Moreland and Beach 1992.

70 Mere exposure: Zajonc 1968.

70 Liking and attractiveness: Kniffin and Wilson 2004.

70 An attractive smile: Rhodes, Sumich, and Byatt 1999.

71 Male or female?: Freud 1933/1965, p. 141.

71 What babies know about males and females: Miller, Younger, and Morse 1982; Quinn et al. 2002.

71 Children's understanding of gender stereotypes: Martin, Eisenbud, and Rose 1995; see Gelman 2003 for review.

72 Island of the boys; island of the girls: Taylor 1996.

72 Interview studies: See Gelman 2003 for review.

72 Cross-dressing in the Bible: Deuteronomy 22:5.

72 Children's disapproval of cross-dressing: Levy, Taylor, and Gelman 1995.

73 Brother-sister incest: Haidt 2001.

73 No worry about sex-crazed siblings: Pinker 1997.

74 Your nakedness: Leviticus 18:10 (translation by Alter 2004).

75 Sexual aversion even if you're not related: Lieberman, Tooby, and Cosmides 2007.

75 What kills the libido?: Lieberman, Tooby, and Cosmides 2007.

76 Men who are wrong about which children share their genes: Anderson 2006.

76 The risk of stepfathers: Daly and Wilson 1999.

77 Babies should look like Dad: Christenfeld and Hill 1995.

77 A bad idea to look like Dad: DeBruine et al. 2008, Pagel 1997.

78 Seats at the official sex table: "Don't ask the sexperts," September 26, 2007, Slate.com: http://www.slate.com/id/2174411.

79 Sham virgins: Cowen 2007.

80 The "virgin cure": Hood 2009.

80 Darwin's list: Desmond and Moore, 1994.

80 Darwin's love: Quammen 2006.

82 The importance of kindness: Buss 1989.

82 Explaining the peacock's tail: Cronin 1991.

83 Sexual selection and the human mind: Miller 2000, quotation from p. 5.

83 Why not a potato?: Miller 2000, p. 124.

84 Why not *Battlestar Galactica*?: Cowen 2007.

85 Evolving a better penis: Miller 2000.

86 The strategic advantage of true love: Pinker 1997, pp. 418, 416.

87 Systems for love and attachment: Fisher 2004.

87 A mother's tie to her baby: Mayes, Swain, and Leckman 2005.

87 The attractions of twins: Wright 1997.

88 Singer's accidental bedtrick: From Pinker 1997, who uses it to make a similar point about how we think about people.

89 Capgras Syndrome: Ramachandran and Blakeslee 1998.

89 A virile and handsome double: Feinberg and Keenan 2004, p. 53. Thanks to Ryan McKay for pointing this out to me.

4. IRREPLACEABLE

91 What money can't buy: Walzer 1984.

92 Taboo trade-offs: Fiske and Tetlock 1997, Tetlock et al. 2000.

92 Hospital administrator: Tetlock et al. 2000.

93 The specialness of money: Ariely 2008.

94 Transaction systems: Fiske 1992; see Pinker 2002 for discussion.

94 Monkey exchanges: Chen, Lakshminaryanan, and Santos 2006.

96 Unused gift cards: For estimates see http://www.consumersunion.org/pub/core_financial_services/005188.html.

98 Endowment effect: Kahneman, Knetsch, and Thaler 1990, 1991.

98 Longer you own it, the more you like it: Strahilevitz and Loewenstein 1998.

98 Wedding gift study: Brehm 1956.

99 Theories of why our choices affect what we like: Bem 1967, Festinger 1957, Lieberman et al. 2001, Steele and Liu 1983.

99 The Brehm effect in children and monkeys: Egan, Santos, and Bloom 2007; Egan, Bloom, and Santos, in press; see also Chen 2008, Chen and Risen 2009, and Sagarin and Skowronski 2009 for discussion.

100 What makes an object valuable?: Frazier, Gelman, and Hood 2009.

100 Kennedy auction: C. McGrath, "A Kennedy plans a tag sale, so Sotheby's expects a crowd," *New York Times*, December 1, 2004.

100 Shakespearean lumber: Pascoe 2005.

101 Napoléon's trees and penis: Pascoe 2007.

101 Foer's collection: Foer 2004.

102 Contagious Magic: Frazer 1922, pp. 37–38.

102 Parts of famous people: Pascoe 2005, p. 3.

103 Sweat and all: Hood 2009.

103 Sweater study: Newman, Diesendruck, and Bloom under review.

104 Touched by an attractive person: Argo, Dahl, and Morales 2008.

104 The destruction of the Wests' home: Hood 2009.

104 Hitler's sweater: Rozin, Millman, and Nemeroff 1986.

105 Specialty auctions for despised objects: G. Stone, "'Murderabilia' sales distress victim's families," August 15, 2007, ABC News Online: http://abcnews.com (search for "murderabilia sales").

106 Berkeley on properties: Berkeley 1713/1979, p. 60.

107 Baby addition: Wynn 1992; see Wynn 2000, 2002 for discussion and Xu 2007 for related studies.

107 Reference to individuals in children's first words: Bloom 2000; see also Macnamara 1982.

107 *Doh?*: Bloom 2000.

107 Duplicating machine studies: Hood and Bloom 2008.

109 Children believe in unusual machines: DeLoache, Miller, and Rosengren 1997.

111 Hamster duplication: Hood and Bloom under review.

112 Story of Bluie: Gopnik 2006, p. 262.

113 Human anthropomorphism: Guthrie 1993.

113 Faces in the moon: Hume 1757/1957, p. 29.

113 Hypertrophy of social intelligence: Boyer 2003, p. 121.

113 Winnicott on transitional objects: Winnicott 1953.

113 Japanese and American children's attachment objects: Hobara 2003.

113 Attachment object experiment: Hood and Bloom 2008.

114 Children's attachment objects: Lehman, Arnold, and Reeves 1995.

115 Billy the attachment object: Hood 2009.

5. PERFORMANCE

117 The Bell experiment: G. Weingarten, "Pearls before breakfast," *Washington Post*, April 8, 2007.

119 Catherine the snob: Koestler 1964, quotations from pp. 403, 408.

120 The story of van Meegeren: Dolnick 2008, Wynne 2006.

121 Greta Garbo: Dutton 2008.

122 Music as a mystery: Darwin 1874/1909.

122 Mekranoti singers: Levitin 2008.

123 Music as proof of God: Vonnegut 2006.

123 Nonhumans don't enjoy music: McDermott and Hauser 2007, Levitin 2008.

123 Babies enjoy music: Trainor and Heinmiller 1998, Trehub 2003.

123 Brain damage can leave you indifferent to music: Sacks 2007.

124 Music as cheesecake: Pinker 1997, quotations from pp. 534, 525.

124 Fiction as a possible adaptation: Pinker 2007.

124 Parallels between music and language: See also Lerdahl and Jackendoff 1983.

125 An adaptationist theory of music: Levitin 2006, 2008.

126 The effects of synchrony: Chartrand and Bargh 1999, Wiltermuth and Heath 2009.

127 Establishing musical tastes in the womb: Lamont 2001.

127 Inverted U: Berlyne 1971.

128 Age and preference for music: Sapolsky 2005, quotation from p. 201; for a related study, see Hargreaves, North, and Tarrant 2006.

128 Musical preferences and social affiliation: Levitin 2006.

129 Avoiding our parents' music: Cowen 2007, p. 67.

129 Experimental aesthetics: See Silva 2006 for review.

130 Pictures as surrogates: Pinker 1997.

130 Monkey porn: Deaner, Khera, and Platt 2005.

131 Naming pictures for the first time: Hochberg and Brooks 1962; see also Ekman and Friesen 1975.

131 Babies understand pictures: DeLoache, Strauss, and Maynard 1979.

131 Grabbing at pictures: DeLoache et al. 1998.

131 Seeing representations as reality: Rozin, Millman, and Nemeroff 1986.

132 The anxiety of cutting up pictures: Hood et al. in press.

132 Throwing darts at pictures of babies: King et al. 2007.

132 Mere exposure and love of impressionist paintings: Cutting 2006.

133 Shape of a painting and its price: Cowen 2007.

133 We care about how art is created: Dutton 2008.

134 Sexual selection and the origin of art: Miller 2000, 2001.

135 For the sake of charming the opposite sex: Darwin 1874/1909, p. 585.

135 People are not peacocks: Hooper and Miller 2008.

136 *Sophie's Choice*: Styron 1979.

137 Nobody thinks about utility: James 1890/1950, p. 386.

138 Art as historic: For instance, Danto 1981; Davies 2004; Levinson 1979, 1989, 1993.

138 Art as performance: Dutton 1983, p. 176.

139 Children's drawings: Bloom 2004, Cox 1992, Winner 1982.

139 Children are sensitive to history when naming art: Bloom 2004, Bloom and Markson 1998, Preissler and Bloom 2008.

140 Benjamin and Malraux on originals: Cited by Kieran 2005.

141 *You* paint a Pollock: Yenawine 1991.

141 Effort and liking: Kruger et al. 2004.

142 The IKEA effect: Norton, Mochon, and Ariely 2009.

142 Marla Olmstead: The events are recounted in the movie *My Child Could Paint That*. See Fineman 2007 for discussion.

143 Several properties that characterize art: Dutton 2008.

143 Pairs of otherwise identical objects: Danto 1981.

143 What do children think is art?: Gelman and Bloom 2000; see also Gelman and Ebeling 1988.

144 Animal art isn't art: Dutton 2008.

144 Rodin's sketches as a counterexample to the performance theory: Kieran 2005.

145 Repeated miscarriages as art: R. Kennedy, "Yale demands end to student's performance," *New York Times*, April 22, 2008.

145 Manzoni's can of feces: Dutton 2008.

146 Hard to store poop in a can: Dutton 2008.

146 From the what of art to the how of art: Menand 2009.

147 *Art* on art: Reza 1997, pp. 3, 15; discussed in Bloom 2004.

147 Accepting the plinth: S. Jones, "Royal Academy's preference for plinth over sculpture leaves artist baffled," *The Guardian*, June 15, 2006.

149 Fuzzy wig, no Einstein: Dolnick 2008, p. 291.

149 Bannister's accomplishment: Gladwell 2001.

149 Speeded waltz: Dutton 1983.

150 Violating the honesty of effort: Gladwell 2001; see also Sandel 2007.

150 Steroids as cheating: Jarudi 2009.

150 Instinctive conservatism: Jarudi 2009; Jarudi, Castaneda, and Bloom under review.

151 The unique and important capacity to share intentions: Tomasello et al. 2005.

152 Fair competes with his mathematical predictions: http://fairmodel.econ.yale.edu/rayfair/marath1.htm.

152 *Guinness Book of World Records* as an illustration of human excellence: Dutton 2008.

153 World Grilled Cheese Eating Competition: Cowen 2007.

153 Problems with beauty: Danto 2007; thanks to Jonathan Gilmore for discussion of these issues.

153 Gurning as a source of aesthetic delight: Kieran 2005.

153 Rules of gurning: http://en.wikipedia.org/wiki/Making_a_face.

6. IMAGINATION

155 Time-management studies: Gleick 2000.

155 Four hours a day in television: From neilsonmedia.com: http://en.us.nielsen.com/main/insights/nielsen_a2m2_three.

155 European obsession with the unreal: Nettle 2005.

156 *Friends* vs. friends: Melanie Green: http://www.unc.edu/~mcgreen/research.html.

156 Imaginative pleasures as a by-product: Pinker 1997.

156 Children's pretending and playing: Harris 2000.

157 Four-year-olds understand pretense: See Skolnick and Bloom 2006a for review.

157 Pretense in babies: Onishi, Baillargeon, and Leslie 2007.

158 Darwin's child: Darwin 1872/1913, p. 358.

158 Play bows: Bekoff 1995.

160 One-year-olds' false belief: Onishi and Baillargeon 2005.

160 Metarepresentation in *Friends*: Zunshine 2006, p. 31.

161 Sending hypotheses ahead: Nuttall 1996, p. 77.

163 Metarepresentation and pretense: Leslie 1994; see also Harris 2000.

164 Chomsky on language: For instance, Chomsky 1987; for a summary, see Pinker 1994.

164 Chinese *Crying Game*: Doniger 2000.

164 Universal themes: McEwan 2005, p. 11; see also Barash and Barash 2008.

165 The importance of human differences: James 1911, p. 256.

165 The puzzle of Anna Karenina: Radford 1975; for other philosophical perspectives, see Gendler and Kovakovich 2005, Morreall 1993, Walton 1990.

166 Psychologists use fiction to study reality: Nichols 2006.

167 Treating a story as fact: Green and Donahue 2009.

168 Robert Young treated as a doctor: Real 1977.

169 Safe but frightened anyway: All examples from Gendler 2008.

169 The notion of alief: Gendler 2008, 2009.

169 What people don't like to do: Rozin, Millman, and Nemeroff 1986; Nemeroff and Rozin 2000 (the gun study is unpublished).

170 Our minds wander when we are left alone: Mason et al. 2007.

170 Lin Yutang quotation: Thanks to Tamar Gendler.

170 Transported into fiction: Gerrig 1993, Green and Brock 2000.

171 Psychological studies of the first person: Reviewed in Coplan 2004.

171 Empathy in *Jaws*: Carroll 1990.

171 Empathy in fiction is like empathy in real life: Coplan 2004.

171 The importance of gossip: Dunbar 1998.

171 The world without us: Weisman 2007.

172 Fiction bustles with intention: Zunshine 2006, p. 26.

172 The evolutionary importance of prestige: Henrich and Gil-White 2001.

173 Fiction is for practicing theory of mind: Zunshine 2006.

173 Fiction is for acquiring social expertise: Mar and Oatley 2008.

173 Fiction is for learning about solutions for real-world problems: Dutton 2008; Pinker 1997, the quotation is from p. 543.

173 Stories motivating moral change: Bloom 2004, Nussbaum 2001.

174 Johnson on the delight of tragedy: Quoted by Nuttall 1996.

176 Conventions for the expression of thought: Zunshine 2008.

176 Voyeurism: McGinn 2005, p. 55.

7. SAFETY AND PAIN

177 Head surgery movies: Haidt, McCauley, and Rozin 1994.

178 Hooptedoodle: E. Leonard, "Easy on the hooptedoodle," *New York Times*, July 16, 2001.

178 To forget words: Wright 2007, p. 280.

178 Best on the big screen: McGinn 2005.

179 Nozick's experience machine: Nozick 1974.

180 To be a character in a story that you don't know is a story: This is part of the plot of *Total Recall*, *The Game*, and *The Truman Show*.

180 Banana peel: See Dale 2000 for an extensive discussion.

182 Canned crying: Jacobs 2004, p. 46.

182 Television has gotten smarter: Johnson 2005.

183 What David Copperfield gets from books: Carroll 2004, p. 48, cited by Dutton 2008.

186 Children are smart about fiction versus reality: See Skolnick and Bloom 2006a for review.

186 Multiple worlds: Skolnick and Bloom 2006b.

187 Smiling makes you happy: Soussignan 2002.

187 Monster in the box: Harris et al. 1991.

188 Adult reluctance to drink imaginary cyanide: Rozin, Markwith, and Ross 2006.

188 Children's understanding of how stories work: Weisberg et al. under review.

190 *Rape Lay*: Alexander 2009.

190 An unaccountable pleasure: Hume 1757/1993, p. 126.

191 The paradox of horror: Carroll 1990.

191 Torture porn: Edelstein 2006.

191 *Blasted*: Patrick Healy, "Audiences gasp at violence; actors must survive it," *New York Times*, November 5, 2008.

192 The problems with catharsis: McCauley 1998.

192 Animal play: Burghardt 2005.

193 Horror movies as safe practice: This proposal is substantially influenced by Denison under review.

193 Tough mind's way of coping: King 1981, p. 316.

194 Plato's gawker: Danto 2003.

194 Benign masochism: Rozin and Vollmecke 1986.

195 Self-injury as a distress call: Hagen under review.

195 Blast of opiates: Berns 2005.

196 Baby punishers: Hamlin, Wynn, and Bloom under review.

196 Altruistic punishment: Fehr and Gächter 2002.

196 Masochism as a form of sadism: Freud 1905/1962.

196 Bad Dobby!: Rowling 2000, p. 12, cited by Nelissen and Zeelenberg 2009.

196 Shocking oneself for sin: Inbar et al. 2008; see also Nelissen and Zeelenberg 2009.

196 Deleuze on masochism: Berns 2005.

197 Elvis has left the building: Bergner 2009.

197 Better magazines: Jerry Seinfeld joke, from Cowen 2007.

197 The sadomasochist who has to go to the dentist: Weinberg, Williams, and Moser, 1984; thanks to Lily Guillot for finding this for me.

197 Half of our waking lives: Klinger 2009.

198 Mind-wandering part of the brain: Mason et al. 2007.

198 Imaginary worlds and multiple selves: Bloom 2008.

198 Children's imaginary companions: Taylor 1999, Taylor and Mannering 2007.

198 Adult fiction writers: Taylor, Hodges, and Kohanyi 2003.

200 Shortage of scarcity: Ainslie 1992, p. 258, cited by Elster 2000.

200 *Twilight Zone* dialogue: http://en.wikipedia.org/wiki/A_Nice_Place_to_Visit.

200 Dreams: Thanks to Marcel Kinsbourne for pointing this out to me.

200 Daydreaming with a friend: Elster 2000.

8. WHY PLEASURE MATTERS

203 Not much food: Fogel 2004, cited by Cowen 2007.

204 Hunter-gatherer appetites in a modern world: Brownell and Horgen 2004.

204 We get pleasure from the natural world: Bloom 2009.

204 The perils of being estranged from nature: Wilson 1999, p. 351.

204 Benefits of nature: See Kahn 1997 for review.

205 Real blood and the pull of magic: Koestler, 1964, p. 405.

206 Essential nonsense: Hood 2009, p. 145.

206 Attachment as irrational: Frazier et al. 2009.

206 Human irrationality: Kahneman, Slovic, and Tversky 1982; for accessible overviews, see Piattelli-Palmarini 1994 and Marcus 2008.

206 Kluges: Marcus 2008.

207 The essence of Pringles: Adam Cohen, "The Lord Justice hath ruled: Pringles are potato chips," *New York Times*, May 31, 2009.

207 Measuring your sanity: Gray 1996.

208 Not smarter, just different: For a similar argument, see Keys and Schwartz 2007.

208 Who's weirder?: Bergner 2009, p. 56.

209 Obsession with luxury goods: Frank 2000, Layard 2005, Miller 2009.

209 Singer on world poverty: For example, Singer 1999, 2009.

210 Escape to nature: Bloom 2009.

211 HDTV study: Kahn, Severson, and Ruckert 2009.

211 Awed wonder: Dawkins 1998, p. x.

212 The minimum definition of religion: Tylor 1871/1958, p. 8.

213 Theories of religious belief: See Bloom 2005, 2007 for reviews.

213 An other reality: Berger 1969, p. 2.

213 Unseen order: James 1902/1994, p. 61.

213 Science shows that the universe is pointless: Weinberg 1977, p. 154, but see Wright 2000 for a critical discussion.

213 Functions of ritual: McCawley and Lawson 2002.

214 The connection between art and religion: Benjamin 2008; Dissanayake 1988, 1992.

214 Children and the ordering of fate: Opie and Opie 1959, p. 210; see Hood 2009 for discussion.

214 Why toddlers are terrible: Gopnik 2000.

214 Child as scientist: Carey 1986, 2009; see also Gopnik 1996.

215 The wonders of Buddhism: Harris 2005, pp. 283–84.

215 Hitchens on the numinous: From his debate with Lorenzo Albacete, September 22, 2008: http://reasonweekly.com (search for "Hitchens Albacete").

216 Triggers to awe: Keltner 2009; see also Keltner and Haidt 2003.

217 Reverence for the collective: Keltner 2009, p. 252.

217 The transforming power of awe: Keltner 2009, p. 252.

218 Orgasm and explanation: Gopnik 2000.

219 Thought experiments: Gendler 2005.

220 Play-acting in religion: Thanks to Peter Gray for discussion about this point.

221 Don't ask the baby: Winnicott 1953, p. 95.

221 A picture of God: Robinson 2009, p. xi.

REFERENCES

Ackerman, J. 2001. *Chance in the house of fate: A natural history of heredity.* New York: Houghton Mifflin.

Aharon, I., Etcoff, N. L., Ariely, D., Chabris, C. F., O'Connor, E., & Breiter, H. C. 2001. Beautiful faces have variable reward value: fMRI and behavioral evidence. *Neuron,* 32:537–51.

Ainslie, G. 1992. *Picoeconomics.* New York: Cambridge University Press.

Alexander, L. 2009. And you thought Grand Theft Auto was bad. Slate.com: http://www.slate.com/id/2213073.

Alter, R. 2004. *The five books of Moses: A translation with commentary.* New York: Norton.

Anderson, K. G. 2006. How well does paternity confidence match actual paternity? *Current Anthropology,* 47:513–20.

Appiah, K. A. 2008. *Experiments in ethics.* Cambridge, MA: Harvard University Press.

de Araujo, I. E., Rolls, E. T., Velazco, M. I., Margot, C., & Cayeux, I. 2005. Cognitive modulation of olfactory processing. *Neuron,* 46:671–79.

Arens, W. 1979. *The man-eating myth: Anthropology and anthropophagy*. New York: Oxford University Press.

Argo, J. J., Dahl, D. W., & Morales, A. C. 2006. Consumer contamination: How consumers react to products touched by others. *Journal of Marketing*, 70:81–94.

———. 2008. Positive consumer contagion: Responses to attractive others in a retail context. *Journal of Marketing Research*, 45:690–712.

Ariely, D. 2008. *Predictably irrational: The hidden forces that shape our decisions*. New York: HarperCollins.

Atran, S. 1998. Folk biology and the anthropology of science: Cognitive universals and cultural particulars. *Behavioral and Brain Science*, 21:547–609.

Baldwin, D. A., Markman, E. M., & Melartin, R. L. 1993. Infants' ability to draw inferences about nonobvious object properties: Evidence from exploratory play. *Cognitive Development*, 64:711–28.

Barash, D. P., & Barash, N. R. 2008. *Madame Bovary's ovaries: A Darwinian look at literature*. New York: Delacorte.

Bartoshuk, L. M., Duffy, V. B., & Miller, I. J. 1994. PTC/PROP tasting: Anatomy, psychophysics, and sex effects. *Physiology and Behavior*, 56:1165–71.

Bekoff, M. 1995. Play signals as punctuation: The structure of social play in canids. *Behaviour*, 132:419–29.

Bem, D. J. 1967. Self-perception: An alternative interpretation of cognitive dissonance phenomena. *Psychological Review*, 74:183–200.

Benjamin, W. 2008. *The work of art in the age of its technological reproducibility, and other writings on media*. Cambridge, MA: Harvard University Press.

Berger, P. L. 1969. *A rumor of angels. Modern society and the rediscovery of the supernatural*. New York: Doubleday.

Bergner, D. 2009. *The other side of desire*. New York: HarperCollins.

Berkeley, G. 1713/1979. *Three dialogues between Hylas and Philonous*. New York: Hackett.

Berlyne, D. E. 1971. *Aesthetics and psychobiology*. New York: Appleton-Century-Crofts.

Berns, G. 2005. *Satisfaction: The science of finding true fulfillment*. New York: Holt.

Birch L. 1999. Development of food preferences. *Annual Review of Nutrition*, 19:41–62.

Bloom, P. 1996. Intention, history, and artifact concepts. *Cognition*, 60:1–29.

———. 1998. Theories of artifact categorization. *Cognition*, 66:87–93.

———. 2000. *How children learn the meanings of words.* Cambridge, MA: MIT Press.

———. 2004. *Descartes' baby: How the science of child development explains what makes us human*. New York: Basic Books.

———. 2005. Is God an accident? *Atlantic Monthly*, December.

———. 2007. Religion is natural. *Developmental Science*, 10:147–51.

———. 2008. First-person plural. *Atlantic Monthly*, November.

———. 2009. Natural happiness. *New York Times Magazine*, April 19.

Bloom, P., & Gelman, S. A. 2008. Psychological essentialism in selecting the 14th Dalai Lama. *Trends in Cognitive Sciences*, 12:243.

Bloom, P., & Markson, L. 1998. Intention and analogy in children's naming of pictorial representations. *Psychological Science*, 9:200–204.

Boese, A. 2007. *Elephants on acid: And other bizarre experiments.* New York: Harvest Books.

Bohannon, J. 2009. Gourmet food, served by dogs. *Science*, 323:1006.

Boyer, P. 2003. Religious thought and behaviour as by-products of brain function. *Trends in Cognitive Sciences*, 7: 119–24.

Brehm, J. W. 1956. Post-decision changes in the desirability of alternatives. *Journal of Abnormal and Social Psychology*, 52:384–89.

Brownell, K. D., & Horgen, K. B. 2004. *Food fight: The inside story of the food industry, America's obesity crisis, and what we can do about it.* New York: McGraw-Hill.

Burghardt, G. M. 2005. *The genesis of animal play: Testing the limits.* Cambridge, MA: MIT Press.

Buss, D. M. 1989. Sex differences in human mate preferences: Evolutionary hypotheses in 37 cultures. *Behavioral and Brain Sciences*, 12:1–49.

Carey, S. 1986. *Conceptual change in childhood.* Cambridge, MA: MIT Press.

———. 2009. *The origin of concepts.* New York: Oxford University Press.

Carroll, J. 2004. *Literary Darwinism: Evolution, human nature, and literature.* New York: Routledge.

Carroll, N. 1990. *The philosophy of horror: Or, paradoxes of the heart*. New York: Routledge.

Chartrand, T. L., & Bargh, J. A. 1999. The chameleon effect: The perception-behavior link and social interaction. *Journal of Personality and Social Psychology*, 76:893–910.

Chen, M. K. 2008. Rationalization and cognitive dissonance: Do choices affect or reflect preferences? Working paper, Yale University, New Haven, CT.

Chen, M. K., Lakshminaryanan, V., & Santos, L. R. 2006. The evolution of our preferences: Evidence from capuchin monkey trading behavior. *Journal of Political Economy*, 114:517–37.

Chen, M. K., & Risen, J. 2009. Is choice a reliable predictor of choice? A comment on Sagarin and Skowronski. *Journal of Experimental Social Psychology* 45:425–27.

Chomsky, N. 1987. *Language and problems of knowledge: The Managua lectures*. Cambridge, MA: MIT Press.

Christenfeld, N. J. S., & Hill, E. A. 1995. Whose baby are you? *Nature*, 378:669.

Clutton-Brock, T. H. 1991. *The evolution of parental care*. Princeton, NJ: Princeton University Press.

Coetzee, J. M. 1995. Meat country. *Granta*, 52:43–52.

Coplan, A. 2004. Empathic engagement with narrative fictions. *Journal of Aesthetics and Art Criticism*, 62:141–52.

Cosmides, L., & Tooby, J. 1994. Beyond intuition and instinct blindness: Towards an evolutionarily rigorous cognitive science. *Cognition*, 50:41–77.

Cowen, T. 2007. *Discover your inner economist*. New York: Penguin.

Cox, M. 1992. *Children's drawings*. London: Penguin Books.

Cronin, H. 1991. *The ant and the peacock*. New York: Cambridge University Press.

Curasi, C. F., Price, L. L., & Arnould, E. J. 2004. How individuals' cherished possessions become families' inalienable wealth. *Journal of Consumer Research*, 11:609–22.

Cutting, J. E. 2006. The mere exposure effect and aesthetic preference. In P. Locher, C. Martindale, L. Dorfman, V. Petrov, & D. Leontiv (Eds.),

New directions in aesthetics, creativity, and the psychology of art. Amityville, NY: Baywood Publishing.

Dale, A. 2000. *Comedy is a man in trouble.* Minneapolis: University of Minnesota Press.

Daly, M., & Wilson, M. 1999. *The truth about Cinderella.* New Haven, CT: Yale University Press.

Danto, A. C. 1981. *The transfiguration of the commonplace.* Cambridge, MA: Harvard University Press.

———. 2003. *The abuse of beauty.* New York: Open Court.

———. 2007. Max Beckmann. In A. Danto (Ed.) *Unnatural wonders: Essays from the gap between art and life.* New York: Columbia University Press.

Darwin, C. 1859/1964. *On the origin of species.* Cambridge, MA: MIT Press.

———. 1872/1913. *The expression of the emotions in man and animals.* New York: D. Appleton.

———. 1874/1909. *The descent of man.* Amherst, NY: Prometheus.

Davies, D. 2004. *Art as performance.* Oxford: Blackwell.

Davies, S. 1991. *Definitions of art.* Ithaca, NY: Cornell University Press.

Dawkins, R. 1998. *Unweaving the rainbow: Science, delusion and the appetite for wonder.* New York: Penguin.

Deaner, R. O., Khera, A. V., & Platt, M. P. 2005. Monkeys pay per view: Adaptive valuation of social images by rhesus macaques. *Current Biology,* 15:543–48.

DeBruine, L. M., Jones, B. C., Little, A. C., & Perrett, D. I. 2008. Social perception of facial resemblance in humans. *Archives of Sexual Behavior,* 37:64–77.

Dekkers, M. 2000. *Dearest Pet: On bestiality.* London: Verso.

DeLoache, J. S., Miller, K. F., & Rosengren, K. S. 1997. The credible shrinking room: Very young children's performance with symbolic and nonsymbolic relations. *Psychological Science,* 8:308–13.

DeLoache, J. S., Pierroutsakos, S. L., Uttal, D. H., Rosengren, K. S., & Gottlieb, A. 1998. Grasping the nature of pictures. *Psychological Science,* 9:205–10.

DeLoache, J. S., Strauss, M., & Maynard, J. 1979. Picture perception in infancy. *Infant Behavior and Development,* 2:77–89.

Denison, R. N. Under review. Emotion practice theory: An evolutionary solution to the paradox of horror.

Dennett, D. C. 1996. *Kinds of minds*. New York: Basic Books.

Desmond, A., & Moore, A. 1994. Darwin: The life of a tormented evolutionist. New York: Norton.

Diamond, J. 1998. *Why is sex fun?* New York: Basic Books.

Diesendruck, G., Gelman, S. A., & Lebowitz, K. 1998. Conceptual and linguistic biases in children's word learning. *Developmental Psychology*, 34:823–39.

Diesendruck, G., Markson, L., & Bloom, P. 2003. Children's reliance on creator's intent in extending names for artifacts. *Psychological Science*, 14:164–68.

Dissanayake, E. 1988. *What is art for?* Seattle: University of Washington Press.

———. 1992. *Homo aestheticus: Where art comes from and why*. New York: Free Press.

Dolnick, E. 2008. *The forger's spell: A true story of Vermeer, Nazis, and the greatest art hoax of the twentieth century*. New York: Harper Perennial.

Doniger, W. 2000. *The bedtrick: Tales of sex and masquerade*. Chicago: University of Chicago Press.

Dunbar, R.I.M. 1998. *Gossip, grooming, and the evolution of language*. Cambridge, MA: Harvard University Press.

Duncan, I.J.H. 2006. The changing concept of animal sentience. *Applied Animal Behavior Science*, 100:11–19.

Dutton, D. 1983. Artistic crimes. In D. Dutton (Ed.), *The forger's art: Forgery and the philosophy of art*. Berkeley and Los Angeles: University of California Press.

———. 2008. *The art instinct: Beauty, pleasure, and human evolution*. New York: Bloomsbury Press.

Edelstein, D. 2006. Now playing at your local multiplex: Torture porn. *New York*, January 28.

Egan, L. C., Bloom, P., & Santos, L. R. In press. Choice-based cognitive dissonance without any real choice: Evidence from a blind two choice paradigm with young children and capuchin monkeys. *Journal of Experimental Social Psychology*.

Egan, L. C., Santos, L. R., & Bloom, P. 2007. The origins of cognitive dissonance: Evidence from children and monkeys. *Psychological Science*, 18:978–83.

Ekman, P., & Friesen, W. V. 1975. *Unmasking the face. A guide to recognizing emotions from facial clues*. Englewood Cliffs, NJ: Prentice-Hall.

Elster, J. 2000. *Ulysses unbound: Studies in rationality, precommitment, and constraints*. New York: Cambridge University Press.

Evans, E. M., Mull, M. A., & Poling, D. A. 2002. The authentic object? A child's-eye view. In S. G. Paris (Ed.), *Perspectives on object-centered learning in museums*. Mahwah, NJ: Lawrence Erlbaum Associates.

Fehr, E., & Gächter, S. 2002. Altruistic punishment in humans. *Nature*, 415:137–40.

Feinberg, T. E., & Keenan, J. P. 2004. Not what, but where, is your "self"? *Cerebrum: The Dana Forum on Brain Science*, 6:49–62.

Festinger, L. 1957. *A theory of cognitive dissonance*. Stanford, CA: Stanford University Press.

Fineman, M. 2007. My kid could paint that. Slate.com: http://www.slate.com/id/2175311.

Fisher, H. 2004. *Why we love: The nature and chemistry of romantic love*. New York: Henry Holt.

Fishman, C. 2007. Message in a bottle. *Fast Company*, December 19.

Fiske, A. P. 1992. The four elementary forms of sociality: Framework for a unified theory of social relations. *Psychological Review*, 99:689–723.

Fiske, A. P., & Tetlock, P. E. 1997. Taboo trade-offs: Reactions to transactions that transgress the spheres of justice. *Political Psychology*, 18:255–97.

Fodor, J. 1988. *Psychosemantics*. Cambridge, MA: MIT Press.

Foer, J. S. 2004. Emptiness. *Playboy*, January, 148–51.

Fogel, R. 2004. *Escape from hunger and premature death, 1700–2100: Europe, America, and the third world*. Cambridge: Cambridge University Press.

Frank, R. H. 2000. *Luxury fever: Money and happiness in an era of excess*. Princeton, NJ: Princeton University Press.

Frazer, J. G. 1922. *The golden bough: A study in magic and religion*. New York: Macmillan.

Frazier, B. N., Gelman, S. A., Wilson, A., & Hood B. 2009. Picasso paintings, moon rocks, and hand-written Beatles lyrics: Adults' evaluations of authentic objects. *Journal of Cognition and Culture*, 9:1–14.

Freud, S. 1905/1962. *Three essays on the theory of sexuality*. Trans. James Strachey. New York: Basic Books.

———. 1933/1965. *New introductory lectures on psycho-analysis*. New York: Norton.

Gelman, S. A. 2003. *The essential child*. New York: Oxford University Press.

Gelman, S. A., & Bloom, P. 2000. Young children are sensitive to how an object was created when deciding what to name it. *Cognition*, 76:91–103.

Gelman, S. A., & Coley, J. D. 1990. The importance of knowing a dodo is a bird: Categories and inferences in 2-year-old children. *Developmental Psychology*, 26:796–804.

Gelman, S. A., & Ebeling, K. S. 1998. Shape and representational status in children's early naming. *Cognition*, 66:835–47.

Gelman, S. A., & Heyman, G. D. 2002. Carrot-eaters and creature-believers: The effects of lexicalization on children's inferences about social categories. *Psychological Science*, 10:489–93.

Gelman, S. A., & Markman, E. M. 1986. Categories and induction in young children. *Cognition*, 23:183–209.

———. 1987. Young children's inductions from natural kinds: The role of categories and appearances. *Child Development*, 58:1532–41.

Gelman, S. A., & Taylor, M. G. 2000. Gender essentialism in cognitive development. In P. H. Miller & E. K. Scholnick (Eds.), *Developmental psychology through the lenses of feminist theories*. New York: Routledge.

Gelman, S. A., & Wellman, H. M. 1991. Insides and essences: Early understandings of the nonobvious. *Cognition*, 38:213–44.

Gendler, T. S. 2005. Thought experiments in science. *Encyclopedia of Philosophy*. New York: Macmillan.

———. 2008. Alief in action (in reaction). *Mind and Language*, 23:552–85.

———. 2009. Alief and belief. *Journal of Philosophy*, 105:634–63.

Gendler, T. S., & Kovakovich, K. 2005. Genuine rational fictional emotions. In M. L. Kieran (Ed.), *Contemporary debates in aesthetics and the philosophy of art*. Oxford: Blackwell.

Gerrig, R. J. 1993. *Experiencing narrative worlds*. New Haven, CT: Yale University Press.

Gil-White, F. J. 2001. Are ethnic groups biological "species" to the human brain? Essentialism in our cognition of some social categories. *Current Anthropology*, 42:515–54.

Gladwell, M. 2001. Drugstore athlete. *The New Yorker*, September 10.

Gleick, J. 2000. *Faster: The acceleration of just about everything*. New York: Vintage.

Gopnik, A. 1996. The scientist as child. *Philosophy of Science*, 63:485–514.

———. 2000. Explanation as orgasm and the drive for causal knowledge: The function, evolution, and phenomenology of the theory formation system. In F. C. Keil & R. A. Wilson (Eds.), *Explanation and cognition*. Cambridge, MA: MIT Press.

Gopnik, A. 2006. *Through the children's gate: A home in New York*. New York: Knopf.

Gould, B. J. 1941. *Discovery, recognition, and installation of the fourteenth Dalai Lama*. New Delhi: Government of India Press. Reprinted in *Discovery, recognition, and enthronement of the fourteenth Dalai Lama: A collection of accounts* (edited by Library of Tibetan Work & Archives). New Delhi: Indraprastha Press.

Gould, S. J. 1989. *Wonderful life: The Burgess shale and the nature of history*. New York: Norton.

Gould, S. J., & Lewontin, R. C. 1979. The spandrels of San Marco and the Panglossian program: A critique of the adaptationist programme. *Proceedings of the Royal Society of London*, 205:281–88.

Graham, S. A., Kilbreath, C. S., & Welder, A. N. 2004. 13-month-olds rely on shared labels and shape similarity for inductive inferences. *Child Development*, 75:409–27.

Gray, P. 1996. What price Camelot? *Time*, May 6.

Green, M. C., & Brock, T. C. 2000. The role of transportation in the persuasiveness of public narratives. *Journal of Personality and Social Psychology*, 78:701–21.

Green, M. C., & Donahue, J. K. 2009. Simulated worlds: Transportation into narratives. In K. D. Markman, W. M. P. Klein, & J. A. Suhr (Eds.),

Handbook of imagination and mental simulation. New York: Psychology Press.

Guthrie, S. E. 1993. *Faces in the clouds: A new theory of religion*. New York: Oxford University Press.

Hagen, E. H. Under review. Gestures of despair and hope: A strategic reinterpretation of deliberate self-harm.

Haidt, J. 2001. The emotional dog and its rational tail: A social intuitionist approach to moral judgment. *Psychological Review*, 108:814–34.

Haidt, J., McCauley, C., & Rozin, P. 1994. Individual differences in sensitivity to disgust: A scale sampling seven domains of disgust elicitors. *Personality and Individual Differences*, 16:701–13.

Hamlin, J., Wynn, K., & Bloom, P. Under review. Third-party reward and punishment in young toddlers.

Hargreaves, D. J., North, A. C., & Tarrant, M. 2006. The development of musical preference and taste in childhood and adolescence. In G. E. McPherson (Ed.), *The child as musician: Musical development from conception to adolescence*. Oxford: Oxford University Press.

Harris, J. R. 1998. *The nurture assumption: Why children turn out the way they do*. New York: Free Press.

Harris, M. 1985. *Good to eat: Riddles of food and culture*. New York: Simon & Schuster.

Harris, P. L. 2000. *The work of the imagination*. Oxford: Blackwell.

Harris, P. L., Brown, E., Marriott, C., Whittall, S., & Harmer, S. 1991. Monsters, ghosts, and witches: Testing the limits of the fantasy-reality distinction in young children. *British Journal of Developmental Psychology*, 9:105–23.

Harris, S. 2005. *The end of faith: Religion, terror, and the future of reason*. New York: Free Press.

Henrich, J., & Gil-White, F. 2001. The evolution of prestige: Freely conferred deference as a mechanism for enhancing the benefits of cultural transmission. *Evolution and Human Behavior*, 22:165–96.

Hirschfeld, L. 1996. *Race in the making: Cognition, culture, and the child's construction of human kinds*. Cambridge, MA: MIT Press.

Hobara, M. 2003. Prevalence of transitional objects in young children in Tokyo and New York. *Infant Mental Health Journal*, 24:174–91.

Hochberg, J., & Brooks, V. 1962. Pictorial recognition as an unlearned ability: A study of one child's performance. *American Journal of Psychology*, 75:624–28.

Hood, B. M. 2009. *SuperSense: Why we believe in the unbelievable*. New York: HarperOne.

Hood, B. M., & Bloom, P. 2008. Children prefer certain individuals over perfect duplicates. *Cognition*, 106:455–62.

———. Under review. Do children believe that duplicating the body also duplicates the mind?

Hood, B. M., Donnelly, K., Leonards, U., & Bloom, P. In press. Modern voodoo: Arousal reveals an implicit belief in sympathetic magic. *Journal of Cognition and Culture*.

Hooper, P. L., & Miller, G. L. 2008. Mutual mate choice can drive costly signaling even under perfect monogamy. *Adaptive Behavior*, 16:53–60.

Hrdy, S. B. 2009. *Mothers and others: The evolutionary origins of mutual understanding*. Cambridge, MA: Harvard University Press.

Hume, D. 1757/1957. *The natural history of religion*. Stanford, CA: Stanford University Press.

———. 1757/1993. Of tragedy. In S. Copley & A. Edgar (Eds.), *Hume: Selected essays*. Oxford: Oxford University Press.

Inagaki, K., & Hatano, G. 2002. *Young children's naive thinking about the biological world*. New York: Psychology Press.

Inbar, Y., Gilovich, T., Pizarro, D., & Ariely, D. 2008. Morality and masochism: Feeling guilt leads to physical self-punishment. Paper presented at the 80th annual meeting of the Midwestern Psychological Association, Chicago, IL.

Jacobs, A. J. 2004. *The know-it-all: One man's humble quest to become the smartest person in the world*. New York: Simon & Schuster.

James W. 1890/1950. *The principles of psychology*. New York: Dover.

———. 1892/1905. *Psychology*. New York: Henry Holt.

———. 1902/1994. *Varieties of religious experience: A study in human nature*. New York: Random House.

———. 1911. *The will to believe: And other essays in popular philosophy*. New York: Longmans, Green, and Co.

Jarudi, I. 2009. Moral psychology is not intuitive moral philosophy. Unpublished doctoral dissertation, Department of Psychology, Yale University.

Jarudi, I., Castaneda, M., & Bloom, P. Under review. Performance enhancement and the status quo bias.

Jaswal, V. K., & Markman, E. M. 2002. Children's acceptance and use of unexpected category labels to draw non-obvious inferences. In W. Gray & C. Schunn (Eds.), *Proceedings of the twenty-fourth annual conference of the Cognitive Science Society*. Mahwah, NJ: Lawrence Erlbaum Associates.

Johnson, C. N., & Jacobs, M. G. 2001. Enchanted objects: How positive connections transform thinking about the very nature of things. Poster presented at the meeting of the Society for Research in Child Development, Minneapolis, MN, April.

Johnson, S. 2005. *Everything bad is good for you: How today's popular culture is actually making us smarter.* New York: Riverhead.

Johnston, V. S., Hagel, R., Franklin, M., Fink, B., & Grammer, K. 2001. Male facial attractiveness: Evidence for hormone-mediated adaptive design. *Evolution and Human Behavior*, 22:251–67.

Jones, B. C., DeBruine, L. M., Perrett, D. I., Little, A. C., Feinberg, D. R., & Law Smith, M. J. 2008. Effects of menstrual cycle phase on face preferences. *Archives of Sexual Behavior*, 37:78–84.

Kahn, P. H., Jr. 1997. Developmental psychology and the biophilia hypothesis: Children's affiliation with nature. *Developmental Review*, 17:1–61.

Kahn, P. H., Jr., Severson, R. L., & Ruckert, J. H. 2009. The human relation with nature and technological nature. *Current Directions in Psychological Science*, 18:37–42.

Kahneman, D., Knetsch, J., & Thaler, R. 1990. Experimental tests of the endowment effect and the Coase theorem. *Journal of Political Economy*, 98:1325–48.

———. 1991. Anomalies: The endowment effect, loss aversion, and status quo bias. *Journal of Economic Perspectives*, 5:193–206.

Kahneman, D., Slovic, P., & Tversky, A. 1982. *Judgment under uncertainty: Heuristics and biases*. New York: Cambridge University Press.

Kass, L. 1992. Organs for sale? Propriety, property, and the price of progress. *The Public Interest*, 107:65–86.

———. 1994. *The hungry soul*. New York: Free Press.

Keil, F. 1989. *Concepts, kinds, and cognitive development*. Cambridge, MA: MIT Press.

Keltner, D. 2009. *Born to be good: The science of a meaningful life*. New York: Norton.

Keltner, D., & Haidt, J. 2003. Approaching awe, a moral, spiritual, and aesthetic emotion. *Cognition and Emotion*, 17:297–314.

Keys, D. J. & Schwartz, B. 2007. "Leaky" rationality: How research on behavioral decision making challenges normative standards of rationality. *Perspectives on Psychological Science*, 2:162–80.

Kieran, M. 2005. *Revealing art*. New York: Routledge.

King, L. A., Burton, C. M., Hicks, J. A., & Drigotas, S. M. 2007. Ghosts, UFOs, and magic: Positive affect and the experiential system. *Journal of Personality and Social Psychology*, 92:905–19.

King, S. 1981. *Danse macabre*. New York: Everest.

Klinger, E. 2009. Daydreaming and fantasizing: Thought flow and motivation. In K. D. Markman, W. M. P. Klein, & J. A. Suhr (Eds.), *Handbook of imagination and mental simulation*. New York: Psychology Press.

Kniffin, K., & Wilson, D. S. 2004. The effect of non-physical traits on the perception of physical attractiveness: Three naturalistic studies. *Evolution and Human Behavior*, 25:88–101.

Koestler, A. 1964. *The act of creation*. New York: Dell.

Kripke, S. 1980. *Naming and necessity*. Cambridge, MA: Harvard University Press.

Kruger, J., Wirtz, D., Van Boven, L., & Altermatt, T. 2004. The effort heuristic. *Journal of Experimental Social Psychology*, 40:91–98.

Lamont, A. M. 2001. Infants' preferences for familiar and unfamiliar music: A socio-cultural study. Paper read at Society for Music Perception and Cognition, August 9.

Langlois, J. H., & Roggman, L. A. 1990. Attractive faces are only average. *Psychological Science*, 1:115–21.

Langlois, J. H., Roggman, L. A., & Rieser-Danner, L. A. 1990. Infants' differential social responses to attractive and unattractive faces. *Developmental Psychology*, 26:153–59.

Langlois, J. H., Kalakanis, L., Rubenstein, A. J., Larson, A., Hallam, M., & Smoot M. 2000. Maxims or myths of beauty? A meta-analytic and theoretical review. *Psychological Bulletin*, 126:390–423.

Layard, R. 2005. *Happiness: Lessons from a new science*. New York: Penguin.

Lee, S., Frederick, D., & Ariely, D. 2006. Try it, you'll like it. The influence of expectation, consumption, and revelation on preferences for beer. *Psychologial Science*, 17:1054–58.

Lehman, E. B., Arnold, B. E., & Reeves, S. L. 1995. Attachments to blankets, teddy bears, and other nonsocial objects: A child's perspective. *Journal of Genetic Psychology*, 156:443–59.

Lehrer, J. 2009. *Proust was a neuroscientist*. New York: Houghton Mifflin.

Lerdahl, F., & Jackendoff, R. 1983. *A generative theory of tonal music*. Cambridge, MA: MIT Press.

Leslie, A. M. 1994. Pretending and believing: Issues in the theory of ToMM. *Cognition*, 50:193–200.

Levinson, J. 1979. Defining art historically. *British Journal of Aesthetics*, 19:232–50.

———. 1989. Refining art historically. *Journal of Aesthetics and Art Criticism*, 47:21–33.

———. 1993. Extending art historically. *Journal of Aesthetics and Art Criticism*, 51:411–23.

Levitin, D. J. 2006. *This is your brain on music*. New York: Dutton.

———. 2008. *The world in six songs: How the musical brain created human nature*. New York: Dutton.

Levy, G. D., Taylor, M. G., & Gelman, S. A. 1995. Traditional and evaluative aspects of flexibility in gender roles, social conventions, moral rules, and physical laws. *Child Development*, 66:515–31.

Lieberman, D., Tooby, J., & Cosmides, L. 2007. The architecture of human kin detection. *Nature*, 445:727–31.

Lieberman, M. D., Ochsner, K. N., Gilbert, D. T., & Schacter, D. L. 2001. Do amnesiacs exhibit cognitive dissonance reduction? The role of explicit memory and attention in attitude change. *Psychological Science*, 12:135–40.

Lindenbaum, S. 2004. Thinking about cannibalism. *Annual Review of Anthropology*, 33:251–69.

Locke, J. 1690/1947. *An essay concerning human understanding*. New York: Dutton.

Macnamara, J. 1982. *Names for things: A study in human learning*. Cambridge, MA: MIT Press.

Mar, R. A., & Oatley, K. 2008. The function of fiction is the abstraction and simulation of social experience. *Perspectives on Psychological Science*, 13:173–92.

Marcus, G. 2008. *Kluge: The haphazard construction of the human mind*. New York: Houghton Mifflin.

Markman, E. 1989. *Categorization and naming in children*. Cambridge, MA: MIT Press.

Martin, C. L., Eisenbud, L., & Rose, H. 1995. Children's gender-based reasoning about toys. *Child Development*, 66:1453–71.

Mason, M. F., Norton, M. I., Van Horn, J. D., Wegner, D. M., Grafton, S. T., & Macrae, C. N. 2007. Wandering minds: The default network and stimulus-independent thought. *Science*, 315:393–95.

Mayes, L. C., Swain, J. E., & Leckman, J. F. 2005. Parental attachment systems: Neural circuits, genes, and experiential contributions to parental engagement. *Clinical Neuroscience Research*, 4: 301–13.

McCauley, C. 1998. When screen violence is not attractive. In J. Goldstein (Ed.), *Why we watch: The attractions of violent entertainment*. New York: Oxford University Press.

McCauley, R. N., & Lawson, E. T. 2002. *Bringing ritual to mind: Psychological foundations of cultural forms*. New York: Cambridge University Press.

McClure, S. M., Li, J., Tomlin, D., Cypert, K .S., Montague, L. M., & Montague, P. R. 2004. Neural correlates of behavioral preference for culturally familiar drinks. *Neuron*, 44:379–87.

McDermott, J., & Hauser, M. D. 2007. Nonhuman primates prefer slow tempos but dislike music overall. *Cognition*, 104:654–68.

McEwan, I. 2005. Literature, science, and human nature. In J. Gottschall & D. S. Wilson (Eds.), *The literary animal: Evolution and the nature of narrative*. Evanston: University of Illinois Press.

McGinn, C. 2005. *The power of movies*. New York: Random House.

McGraw, A. P., Tetlock, P. E., & Kristel, O. V. 2003. The limits of fungibility: Relational schemata and the value of things. *Journal of Consumer Research*, 30:219–29.

McLaren, A. 2007. *Impotence: A cultural history*. Chicago: University of Chicago Press.

Medin, D. L. 1989. Concepts and conceptual structure. *American Psychologist*, 44:1469–81.

Medin, D. L., & Ortony, A. 1989. Psychological essentialism. In S. Vosniadou & A. Ortony (Eds.), *Similarity and analogical reasoning*. New York: Cambridge University Press.

Menand, L. 2002. What comes naturally. *The New Yorker*, November 22.

———. 2009. Saved from drowning: Barthelme reconsidered. *The New Yorker*, February 23.

Michaels, J. 2007. Three selections from *The masochist's cookbook*. http://www.mcsweeneys.net/2007/6/5michaels.html.

Miller, C. L., Younger, B. A., & Morse, P. A. 1982. The categorization of male and female voices in infancy. *Infant Behavior and Development*, 5:143–59.

Miller, G. F. 2000. *The mating mind: How sexual selection shaped human nature*. London: Heinemann.

———. 2001. Aesthetic fitness: How sexual selection shaped artistic virtuosity as a fitness indicator and aesthetic preferences as mate choice criteria. *Bulletin of Psychology and the Arts*, 2:20–25.

———. 2009. *Spent: Sex, evolution, and consumer behavior*. New York: Viking.

Miller, G. F., Tybur, J., & Jordan, B. 2007. Ovulatory cycle effects on tip earnings by lap-dancers: Economic evidence for human estrus? *Evolution and Human Behavior*, 28:375–81.

Miller, W. I. 1997. *The anatomy of disgust*. Cambridge, MA: Harvard University Press.

———. 1998. Sheep, joking, cloning, and the uncanny. In M. C. Nussbaum & C. R. Sunstein (Eds.), *Clones and clones: Facts and fantasies about human cloning*. New York: Norton.

Moreland, R. D., & Beach, S. 1992. Exposure effects in the classroom: The development of affinity among students. *Journal of Experimental Social Psychology*, 28:255–76.

Morreall, J. 1993. Fear without belief. *The Journal of Philosophy*, 90:359–66.

Nelissen, R. M., & Zeelenberg, M. 2009. When guilt evokes self-punishment: Evidence for the existence of a Dobby effect. *Emotion*, 9: 118–22.

Nemeroff, C., & Rozin, P. 1989. "You are what you eat": Applying the demand-free "impressions" technique to an unacknowledged belief. *Ethos: The Journal of Psychological Anthropology*, 17:50–69.

———. 2000. The makings of the magical mind. In K. S. Rosengren, C. N. Johnson, & P. L. Harris (Eds.), *Imagining the impossible: Magical, scientific, and religious thinking in children*. New York: Cambridge University Press.

Nettle, D. 2005. What happens in *Hamlet*? Exploring the psychological foundation of drama. In J. Gottschall & D. S. Wilson (Eds.), *The literary animal: Evolution and the nature of narrative*. Evanston: University of Illinois Press.

Newman, G., Diesendruck, G., & Bloom, P. Under review. Celebrity contagion and the value of objects.

Nichols, S. 2006. Introduction. In S. Nichols (Ed.), *The architecture of the imagination: New essays in pretence, possibility, and fiction*. New York: Oxford University Press.

Norton, M. I., Mochon, D., & Ariely, D. 2009. *The IKEA effect: Why labor leads to love*. Paper presented at the Society of Personality and Social Psychology, Tampa, FL.

Nozick, R. 1974. *Anarchy, state, and utopia*. New York: Basic Books.

Nussbaum, M. C. 2001. *Upheavals of thought: The intelligence of emotions*. New York: Cambridge University Press.

Nuttall, A. D. 1996. *Why does tragedy give pleasure?* New York: Oxford University Press.

Onishi, K. H., & Baillargeon, R. 2005. Do 15-month-old infants understand false beliefs? *Science*, 308:255–58.

Onishi, K. H., Baillargeon, R., & Leslie, A. M. 2007. 15-month-old infants detect violations in pretend scenarios. *Acta Psychologica*, 124:106–28.

Opie, I., & Opie, P. 1959. *The lore and language of schoolchildren*. New York: Oxford University Press.

Pagel, M. 1997. Desperately concealing father: A theory of parent-infant resemblance. *Animal Behaviour*, 53:973–81.

Pascoe, J. 2005. *The hummingbird cabinet: A rare and curious history of romantic collectors.* Ithaca, NY: Cornell University Press.

———. 2007. Collect-me-nots. *New York Times*, May 17.

Penton-Voak, I. S., Perrett, D. I., Castles, D., Burt, M., Koyabashi, T., & Murray, L. K. 1999. Female preference for male faces changes cyclically. *Nature*, 399:741–42.

Perrett, D. I., May, K. A., Yoshikawa, S. 1994. Facial shape and judgments of female attractiveness. *Nature*, 368:239–42.

Piattelli-Palmarini, M. 1994. *Inevitable illusions: How mistakes of reason rule our minds.* New York: Wiley.

Pinker, S. 1994. *The language instinct.* New York: Norton.

———. 1997. *How the mind works.* New York: Norton.

———. 2002. *The blank slate: The denial of human nature in modern intellectual life.* New York: Viking.

———. 2007. Toward a consilient study of literature. *Philosophy and Literature*, 31:161–77.

———. 2008. The moral instinct. *New York Times Magazine*, January 13.

Plassmann, H., O'Doherty, J., Shiv, B., & Rangel, A. 2008. Marketing actions can modulate neural representations of experienced pleasantness. *Proceedings of the National Academy of Sciences*, 105:1050–54.

Pollan, M. 2006. *The omnivore's dilemma: A natural history of four meals.* New York: Penguin.

Preissler, M. A., & Bloom, P. 2008. Two-year-olds use artist intention to understand drawings. *Cognition*, 106:512–18.

Putnam, H. 1973. Meaning and reference. *Journal of Philosophy*, 70:699–711.

———. 1975. The meaning of "meaning." In H. Putnam (Ed.), *Philosophical papers 2: Mind, language and reality.* Cambridge: Cambridge University Press.

Quammen, D. 2006. *The reluctant Mr. Darwin: An intimate portrait of Charles Darwin and the making of his theory of evolution.* New York: Norton.

Quinn, P. C., Yahr, J., Kuhn, A., Slater, A. M., & Pascalis, O. 2002. Representation of the gender of human faces by infants: A preference for female. *Perception*, 31:1109–21.

Radford, C. 1975. How can we be moved by the fate of Anna Karenina? *Proceedings of the Aristotelian Society*, 49:67–80.

Ramachandran, V. S., & Blakeslee, S. 1998. *Phantoms in the brain.* New York: Harper Perennial.

Rawson, C. 1985. Eating people. *London Review of Books*, January 24.

Real, M. R. 1977. *Mass-mediated culture.* Edgewood Cliffs, NJ: Prentice-Hall.

Reza, Y. 1997. *Art: A play.* Trans. C. Hampton. London: Faber & Faber.

Rhodes, G. 2006. The evolutionary psychology of facial beauty. *Annual Review of Psychology*, 57:199–226.

Rhodes, G., Sumich, A., & Byatt, G. 1999. Are average facial configurations attractive only because of their symmetry? *Psychological Science*, 10:52–58.

Robinson, K. 2009. *The element: How finding your passion changes everything.* New York: Viking.

Rowling, J. K. 2000. *Harry Potter and the chamber of secrets.* New York: Scholastic.

Rozin, P. 1976. The selection of food by rats, humans, and other animals. In J. S. Rosenblatt, R. A. Hinde, E. Shaw, & C. Beer (Eds.), *Advances in the study of behavior, vol. 6.* New York: Academic Press.

———. 1986. One-trial acquired likes and dislikes in humans: Disgust as a US, food predominance, and negative learning predominance. *Learning and Motivation*, 17:180–189.

———. 2004. Meat. In S. Katz (Ed.), *Encyclopedia of food.* New York: Scribner.

———. 2005. The meaning of "natural": Process more important than content. *Psychological Science*, 16:652–58.

———. 2006. Domain denigration and process preference in academic psychology. *Perspectives on Psychological Science*, 1:365–76.

Rozin, P., & Fallon, A. 1987. A perspective on disgust. *Psychological Review*, 94:23–41.

Rozin, P., Haidt, J., & McCauley, C. R. 2000. Disgust. In M. Lewis & J. M. Haviland-Jones (Eds.), *Handbook of emotions, 2nd ed.* New York: Guilford Press.

Rozin, P., Markwith, M., & Ross, B. 2006. The sympathetic magical law of similarity, nominal realism, and neglect of negatives in response to negative labels. *Psychological Science*, 1:383–84.

Rozin, P., Millman, L., & Nemeroff, C. 1986. Operation of the laws of sympathetic magic in disgust and other domains. *Journal of Personality and Social Psychology*, 50:703–12.

Rozin, P., & Schiller, D. 1980. The nature and acquisition of a preference for chili pepper by humans. *Motivation and Emotion*, 4:77–101.

Rozin, P., & Vollmecke, T. A. 1986. Food likes and dislikes. *Annual Review of Nutrition*, 6:433–56.

Sacks, O. 2007. *Musicophilia: Tales of music and the brain*. New York: Knopf.

Sagarin, B. J., & Skowronski, J. J. 2009. The implications of imperfect measurement for free-choice carry-over effects: Reply to M. Keith Chen's (2008) "Rationalization and cognitive dissonance: Do choices affect or reflect preferences?" *Journal of Experimental Social Psychology*, 45:421–23.

Salinger, J. D. 1959. *Raise high the roof beam, carpenters, and Seymour: An introduction*. New York: Little, Brown & Company.

Sandel, M. J. 2007. *The case against perfection: Ethics in the age of genetic engineering*. Cambridge, MA: Harvard University Press.

Sapolsky, R. M. 2005. *Monkeyluv: And other essays on our lives as animals*. New York: Scribner.

Sendak, M. 1988. *Where the wild things are*. New York: HarperCollins.

Shutts, K., Kinzler, K. D., McKee, C. B., & Spelke, E. S. 2009. Social information guides infants' selection of foods. *Journal of Cognition and Development*, 10:1–17.

Siegal, M., & Share, D. 1990. Contamination sensitivity in young children. *Developmental Psychology*, 26:455–58.

Silva, P. J. 2006. *Exploring the psychology of interest*. New York: Oxford University Press.

Singer, P. 1999. The Singer solution to world poverty. *New York Times Magazine*, September 5.

———. 2009. *The life you can save: Acting now to end world poverty*. New York: Random House.

Skolnick, D., & Bloom, P. 2006a. The intuitive cosmology of fictional worlds. In S. Nichols (Ed.), *The architecture of the imagination: New essays on pretense, possibility, and fiction*. Oxford: Oxford University Press.

———. 2006b. What does Batman think about SpongeBob? Children's understanding of the fantasy/fantasy distinction. *Cognition*, 101:B9–B18.

Slater, A., Von der Schulenburg, C., Brown, E., Badenoch, M., Butterworth, G., Parsons, S., & Samuels, C. 1998. Newborn infants prefer attractive faces. *Infant Behavior and Development*, 21:345–54.

Smith, E. W. 1961. The power of dissonance techniques to change attitudes. *Public Opinion Quarterly*, 25:626–39.

Smith, J. 1995. People eaters. *Granta*, 52:69–84.

Smith, J., & Russell, G. 1984. Why do males and females differ? Children's beliefs about sex differences. *Sex Roles*, 11:1111–20.

Soussignan, R. 2002. Duchenne smile, emotional experience, and autonomic reactivity: A test of the facial feedback hypothesis. *Emotion*, 2:52–74.

Steele, C. M., & Liu, T. J. 1983. Dissonance processes as self-affirmation. *Journal of Personality and Social Psychology*, 45:5–19.

Strahilevitz, M., & Lowenstein, G. 1998. The effect of ownership history on the valuation of objects. *Journal of Consumer Research*, 25:276–89.

Styron, W. 1979. *Sophie's choice*. New York: Random House.

Sylvia, C., & Nowak, W. 1997. *A change of heart: A memoir*. New York: Time Warner.

Tajfel, H. 1970. Experiments in intergroup discrimination. *Scientific American*, 223:96–102.

———. 1982. Social psychology of intergroup relations. *Annual Review of Psychology*, 33:1–39.

Taylor, M. 1996. The development of children's beliefs about social and biological aspects of gender differences. *Child Development*, 67:1555–71.

———. 1999. *Imaginary companions and the children who create them*. New York: Oxford University Press.

Taylor, M., Hodges, S. D., & Kohanyi, A. 2003. The illusion of independent agency: Do adult fiction writers experience their characters as having minds of their own? *Imagination, Cognition, and Personality*, 22:361–80.

Taylor, M., & Mannering, A. M. 2007. Of Hobbes and Harvey: The imaginary companions of children and adults. In A. Goncu & S. Gaskins (Eds.), *Play and development: Evolutionary, sociocultural and functional perspectives*. Mahwah, NJ: Lawrence Erlbaum Associates.

Taylor, T. 2004. *The buried soul: How humans invented death*. Boston: Beacon Press.

Tetlock, P. E., Kristel, O. V., Elson, B., Green, M. C., & Lerner, J. 2000. The psychology of the unthinkable: Taboo trade-offs, forbidden base rates, and heretical counterfactuals. *Journal of Personality and Social Psychology*, 78:853–70.

Theroux, P. 1992. *The happy isles of Oceania*. New York: Putnam.

Tomasello, M., Carpenter, M., Call, J., Behne, T., & Moll, H. 2005. Understanding and sharing intentions: The origins of cultural cognition. *Behavioral and Brain Sciences*, 28:675–91.

Trainor, L. J., & Heinmiller, B. M. 1998. The development of evaluative responses to music: Infants prefer to listen to consonance over dissonance. *Infant Behavior and Development*, 21:77–88.

Trehub, S. E. 2003. The developmental origins of musicality. *Nature Neuroscience*, 6:669–73.

Trivers, R. L. 1972. Parental investment and sexual selection. In B. Campbell (Ed.), *Sexual selection and the descent of man, 1871–1971*. Chicago: Aldine.

Tylor, E. B. 1871/1958. *Primitive culture, vol 2: Religion in primitive culture*. New York: Harper & Brothers.

Vonnegut, K. 2006. Vonnegut's blues for America. *Sunday Herald* (Scotland), February 5.

Walton, K. L. 1990. *Mimesis as make-believe*. Cambridge, MA: Harvard University Press.

Walzer, M. 1984. *Spheres of justice: A defense of pluralism and equality*. New York: Basic Books.

Wangdu, K. S. 1941. Report on the discovery, recognition, and enthronement of the fourteenth Dalai Lama. New Delhi: Government of India Press. Reprinted in *Discovery, recognition, and enthronement of the fourteenth Dalai Lama: A collection of accounts* (edited by Library of Tibetan Work & Archives). New Delhi: Indraprastha Press.

Wedekind, C., & Füri, S. 1997. Body odour preferences in men and women: Do they aim for specific MHC combinations or simply heterozygosity? *Proceedings of the Royal Society of London, Series B, Biological Sciences*, 264:1471–79.

Weinberg, M. S., Williams, C. J., & Moser, C. 1984. The social constituents of sadomasochism. *Social Problems*, 31:379–89.

Weinberg, S. 1977. *The first three minutes: A modern view of the origin of the universe*. New York: Basic Books.

Weisberg, D. S., Sobel, D. M., Goodstein, J., & Bloom, P. Under review. Preschoolers are reality-prone when constructing stories.

Weisman, A. 2007. *The world without us*. New York: Thomas Dunne Books.

Welder, A. N., & Graham, S. A. 2001. The influence of shape similarity and shared labels on infants' inductive inferences about nonobvious object properties. *Child Development*, 72:1653–73.

Wilson, E. O. 1999. *The diversity of life*. New York: Norton.

Wiltermuth, S. S., & Heath, C. 2009. Synchrony and cooperation. *Psychological Science*, 20:1–5.

Winner, E. 1982. *Invented worlds: The psychology of the arts*. Cambridge, MA: Harvard University Press.

Winnicott, D. W. 1953. Transitional objects and transitional phenomena: A study of the first not-me possession. *International Journal of Psychoanalysis*, 34:89–97.

Wright, L. 1997. *Twins: And what they tell us about who we are*. New York: Wiley.

Wright, R. 2000. *Nonzero: The logic of human destiny*. New York: Little, Brown.

Wright, R. N. 2007. *Black boy: A record of childhood and youth*. New York: Harper Perennial.

Wynn, K. 1992. Addition and subtraction by human infants. *Nature*, 358:749–50.

———. 2000. Findings of addition and subtraction in infants are robust and consistent: A reply to Wakeley, Rivera and Langer. *Child Development*, 71:1535–36.

———. 2002. Do infants have numerical expectations or just perceptual preferences? *Developmental Science*, 2:207–9.

Wynne, F. 2006. *I was Vermeer: The rise and fall of the twentieth century's greatest forger*. New York: Bloomsbury.

Xu, F. 2007. Sortal concepts, object individuation, and language. *Trends in Cognitive Sciences*, 11:400–406.

Yenawine, P. 1991. *How to look at modern art*. New York: Harry N. Abrams.

Zajonc, R. B. 1968. Attitudinal effects of mere exposure. *Journal of Personality and Social Psychology Monographs*, 9:1–27.

Zunshine, L. 2006. *Why we read fiction: Theory of mind and the novel*. Columbus: Ohio State University Press.

———. 2008. Theory of mind and fictions of embodied transparency. *Narrative*, 16:65–92.

INDEX

Act of Creation, The (Koestler), 119–20
adaptationist theories, xiii, 5, 6, 8, 12, 58, 60, 64, 69, 85, 156
  of awed emotional responses, 217–18
  evolutionary theories vs., xiii
  of imaginative abilities and pleasures, 161–63, 171–73
  of music pleasure, 124–26
  aesthetics, experimental, 129
AIDS, 80
Ainslie, George, 200
albinos, 41
Ali, Muhammad, 149
alief, 169–71, 184
  as more intense in children, 186–88
Allen, Melissa, xv, 139
Almodóvar, Pedro, 199
altruistic punishment, 196
Alzheimer's disease, 204
Amazing Race, The (TV show), 176
Amazon, 122
Anatomy of Disgust, The (Miller), 34
animals, 4, 5, 6, 23, 123, 203–4
  art creation by, 144
  aversion to eating of, 30, 33, 35, 53
  food habits of, 31, 36, 50
  imaginative abilities in, 158, 159, 170, 192–93
  object valuation in, 99
  sexual reproduction habits of, 57–61, 62, 67, 82–83, 85, 134, 135
  visual pleasure displayed by, 130
Aniston, Jennifer, 172
anthropomorphization, 113
aphrodisiacs, 41, 44
Appiah, Kwame Anthony, 53
Ardrey, Robert, 5
Ariely, Dan, 93, 94
Aristotle, 192, 219–20
Armstrong, Neil, 3

Art (play), 147
Art Instinct, The (Dutton), 133–34
artwork, xi, xii, xiv, 1–3, 5, 7, 22, 58, 112, 130, 154, 182, 214, 219
  context in appreciation of, 118, 119
  creative intention in appreciation of, 138–39, 143–45, 146, 210
  as "Darwinian fitness tests," 133–37
  enhancing of status with, 133
  evolutionary theories of appreciation of, 133–37
  forgeries and fakes in, 2, 108, 119, 120–22, 140, 148, 149, 206
  "mere exposure" effect in appreciation of, 132–33
  modern, 43, 120, 133, 140–41, 146, 153
  originals holding special value in, 3, 87, 119–20, 140, 206
  origins and history in appreciation of, 2–3, 87, 119–22, 138–47
  perceived effort in appreciation of, 141–42
  performance theory for defining of, 142–47
  and pleasure in virtuoso displays, 112, 137–38, 140–41, 142, 146, 149, 154, 179, 183, 209
  positive contagion in appreciation of, 133, 140, 145
  sexual selection theory of origin of, 133–36
  signaling theory in purchasing of, 43, 140
  snobbery in appreciation of, 119, 120, 122
  sports vs., 148–49, 151
  ugliness displays in, 153
  see also paintings
atheists, 215
attachment objects, xi, xii, 113–15, 206, 208
Australia, 134
autism, 11, 157, 204
averageness, physical attractiveness and, 65, 66, 69, 70
awe, emotion of, 212, 216–18
Axe, 19
Aztecs, 38

babies, 65, 69, 71, 105, 129, 131
  distinguishing of pretense and reality in, 157–58
  essentialist mindset in, 15, 111
  food preferences in, 32, 34
  music appreciation in, 123, 124, 127
  reasoning about individual objects in, 106–7
  "transitional objects" of, 113, 220–21
  see also children
Bannister, Roger, 149
Bartoshuk, Linda, 28
bedtricks, 55–57, 88, 164
beer, 47–48
Bell, Joshua, 117–18, 119, 122, 135
benign masochism, 51–52, 194–95, 196
Benjamin, Walter, 140, 214
Berger, Peter, 213
Bergner, Daniel, 67, 197, 208
Berkeley, George, 106
Bhagavad Gita, 216
biological essentialism, 10, 12, 13, 15, 17, 22, 106, 151, 207, 211–12, 213, 214
  art appreciation and, 134–35, 136–37
  in gender differences, 15, 17, 57–65, 72
  physical attraction and, 65–67, 86
  in sexuality and reproduction, 57–67, 74, 76, 77, 83, 134–35
  see also evolutionary theories

Blasted (play), 191–92
blind taste tests, 44, 45, 47
blog wars, 204
Bloom, Max, xvi, 139
Bloom, Zachary, xvi, 188
Boijmans Gallery, 120
Bordeaux wine, 45
Borges, Jorge Luis, 10–11
Bosch, Hieronymus, 153
Botox, 68, 210
bottled water, 42–43, 44, 49, 50, 83, 210
bowerbirds, 134, 135
Brandes, Bernd, 25, 26, 27, 53
Brazil, 38
Brehm, Jack, 98
Brillo Box (Warhol), 144
British Broadcasting Standards Commission, 40
Brochette, Frederic, 46
Brooks, Virginia, 131
Bruce Museum, 121
Brudos, Jerome, 67
Buddhism, 215
Budweiser, 47
Buffy the Vampire Slayer (TV show), 164
bugs, 30, 33, 34, 35
Burke, Edmund, 216
Bush, George W., 3

Cage, John, 144, 145
Canada, 30
cannibalism, 25–27, 29, 32–33, 34, 35–36, 44, 53
  eating placentas as form of, 40
  human body part trafficking and, 40–41
  life-force essentialism and, 20, 26–27, 36–41, 206–7
  in religious ritual, 39
  two types of, 36–38
Capgras Syndrome, 89
capuchin monkeys, 99
Carey, Susan, xv, 214
Carroll, Joseph, 183
Carroll, Noël, 171, 191
category essentialism, 10–18, 22, 30–31, 207
  female virginity and, 77–80, 206, 209
  food aversions and, 30, 33–35
  gender differences and, 14, 15, 17, 59, 71–73
  life-force vs., 20–21
  in sexual preferences, 71–80
catharsis, 192
Catholics, Catholicism, 39, 214, 220
celebrity memorabilia, xii, 3, 19, 99–101, 102, 103–5, 108, 110, 112, 115, 207
Celestial Emporium of Benevolent Knowledge, The (Borges), 10–11
Chaplin, Charlie, 181
Charles, Prince of Wales, 20
cheating, 149–51
children, 41, 63, 64, 80, 95, 125, 136–37, 167, 176, 185, 196, 214–15
  ability to distinguish pretense and reality in, 157–58, 163, 168, 185–86, 187
  alief more intensely experienced in, 186–88
  appreciation of realistic images in, 131
  art appreciation and creation in, 136, 143–44
  attachment objects of, xi, 113–15, 206, 208, 220–21
  emergence of disgust in, 34–35
  essentialist mindset in, 14–18, 72, 93, 110–12, 114, 138–39, 208, 214, 215
  food preferences in, 34, 45

children (*continued*)
  gender differences as perceived by, 14, 17, 71–73
  imaginative abilities in, xi, 156–57, 158–59, 163, 168, 185–86, 187, 192–93, 198, 220, 221
  object valuation by, xi, 87, 93, 99, 105–10, 113–15, 206, 208, 220–21
  performance tendencies in, 151–52
  pretend play in, 156–57, 158, 192–93, 220
  see also babies
chimpanzees, 5, 36, 123, 144, 164
China, 75
Chomsky, Noam, 164
Christianity, 39, 78, 212, 214, 220
Christ with the Woman Taken in Adultery (van Meegeren), 1
claques, 182
Clinton, Bill, 78
Clooney, George, 19, 103, 105, 115
Coen brothers, 199
Cohen, Emma, xv, 19
Coke, 45, 48, 50
Communal Sharing, 94, 96
competition, 152–54
"conspicuous consumption," 43
Corcoran School, 118
Cosmides, Leda, 75
costly signaling, 83–84
Cowen, Tyler, 84, 128–29
cows, 30
Crying Game, The, 56, 164
cultural theories, 28, 59
  evolutionary vs., 6–8
  of food preferences, 29–32, 34–35, 53
  of gender differences, 72
  of sexual pleasure, 59
Cutting, James, 132–33

Dahmer, Jeffrey, 26–27, 104, 105
Dalai Lama, 21–22, 218
darshan, 20
Darwin, Charles, 33, 34, 60, 65, 80–82, 122, 134, 135, 158, 163
Darwin, William, 158
Da Vinci Code, The (Brown), 168
Dawkins, Richard, 74, 215
daydreaming, 155, 163, 170, 178, 193, 197–201
  mashochistic, 156, 199, 200
  shortfalls of, 199–200
Deleuze, Gilles, 196
Dennett, Daniel, 18
Descartes, René, 179
Descartes' Baby (Bloom), 34
Deuteronomy, 72
Diamond, Jared, 62
Diana, Princess of Wales, 3, 172
Dickens, Charles, 165, 183
Diesendruck, Gil, xv, 103
disgust:
  developmental emergence of, 34–35
  food preferences shaped by, 32–35
Dissanayake, Ellen, 214
Disturbia, 176
dogs, 5, 30, 50, 123, 158, 170
Doniger, Wendy, 55, 59
Donkey Kong, 153
Duchamp, Marcel, 143, 145, 148, 153
Dutton, Denis, xiv, xv, 133–34, 135, 136, 138, 143, 149, 152, 173
Dylan, Natalie, 79

eBay, 100
Egan, Louisa, xv, 99
Einstein, Albert, 100, 148–49
elephants, 5, 144
Elizabeth II, Queen of England, 108, 110

Elster, Jon, 200
Emin, Tracey, 143
empathy, 171
endocannibalism, 36–37
endowment effect, 98–99
Equity Matching, 94, 96
Essential Child, The (Gelman), 14
essentialism, essences, 8–24, 129, 205–17
  and appeal of nature and purity, 42–43, 44, 49, 53, 68, 151, 210–11
  in art and performance, *see* artwork; performance valuation
  awed emotional responses and, 212, 216–18
  bias and flaws in, 12–14, 49–50, 206–7
  biological, *see* biological essentialism; evolutionary theories
  category, *see* category essentialism
  children as possessing concept of, 14–18, 72, 110–12, 114, 138–39, 208, 214, 215
  contacting the transcendent and, 23, 212–15, 216, 221
  creative intention and, 10, 17, 138–39, 143–45, 146, 210
  curiosity in and seeking out of, 10, 210–15, 218, 221
  definition of, xii, 9
  in everyday objects, *see* object valuation
  in food pleasures, *see* food pleasures
  history and origins as factor in, 2–4, 10, 18–19, 93, 99–105, 115, 119–22, 138–47, 154, 205, 207, 208
  human pleasures as by-product of, 8–9, 22
  in imaginative pleasures, *see* imagination and imaginative pleasures
  immoral acts driven by, 209
  life-force, *see* life-force essentialism
  money taboos and, 93–97
  perceived irrationality of, 119, 205–9
  as pervasive in language, 10–12, 16–17
  in religious ritual, 19, 21–22, 39, 212–14
  science in uncovering of, 10, 12, 15, 211–12, 214, 215
  in sexuality and sexual pleasures, *see* sexuality and sexual pleasures
  signaling theory vs., 43–44
  of social individuals, 110–12, 207
  theories of origin of, xiii, 15
  in virtuoso displays, 112, 137–38, 140–41, 142, 146, 149–50, 153, 154, 179–83, 209
etiquette, violations of, 52–53
Eucharist, 39, 214, 220
Everything Bad Is Good for You (Johnson), 182–83
evolutionary theories, xiii–xiv, 36, 49, 82, 105, 151, 171, 203–5, 206
  adaptationist theories vs., xiii
  of art origins and appreciation, 133–37
  of awed emotional responses, 217–18
  cultural and social theories vs., 6–8
  food preferences and, 27–28
  of function of pleasure, 4–8
  of gender differences, 59–65
  of imaginative abilities and pleasures, xiii–xiv, 161–63, 171–73, 201
  of music appreciation, 122–26, 135
  of sexual pleasure and reproduction, 57–65, 69, 74, 77, 79, 82–85
  of sexual selection, 63–64, 74, 82–85, 86, 133–36
  see also adaptationist theories

exocannibalism, 37–38
experimental aesthetics, 129
"Explanation as Orgasm" (Alison Gopnik), 218

Fair, Ray C., xv, 152
Family Guy, 49
Fear Factor (TV show), 176
female orgasm, 58
female virginity, xi, 77–80, 206, 209
fetishes, sexual, 58, 67–68
fiction, xi, xii, xiii–xiv, 124, 152, 156, 163–68, 170, 171–76, 178–92, 196, 198–99, 219
  alief experienced with, 184, 186–88
  appreciating virtuosity of creator in, 179–83, 209
  catharsis in, 192
  confounding reality with, 167–68, 187
  emotional triggering by, 165–68, 171, 174–75, 184, 186–88, 190–92
  evolving tastes in, 182–83
  humor in, 180–81, 182
  as more compelling than reality, 175–76
  pain and fears explored in, 178, 190, 191–92, 193–94
  as practice for dealing with real-life horrors, 173, 193–94
  psychic intimacy in, 175–76, 184
  religious role of, 220
  safety in, 173, 183–85, 188–89, 190, 193
  as social learning tool, 171–73
  technology and increasing realism in, 178–79
  tragedy in, 174–75, 190, 191, 192
  universal plots in, 7, 164–65
  violence and horror as appealing in, 181, 189, 190–92, 193–94
Fiske, Alan, 94
Fodor, Jerry, 15
Foer, Jonathan Safran, 101
food pleasures, xi–xii, 5, 6, 7, 8, 25–53, 128, 137, 209
  and aversions, 30, 31, 32–35, 52
  biological role in, 27–29
  cannibalism and, 20, 25–27, 29, 32–33, 34, 35–41, 44, 53, 206–7
  cultural theories of, 29–32, 34–35, 53
  etiquette violations and, 52–53
  inverted U rule of, 127
  life-force essentialism in, 20, 26–27, 36–41, 206–7
  morality in, 53
  in natural and pure goods, 42–43, 44, 49, 53
  perception of taste as colored by beliefs in, 44–50
  personal experience and observation in shaping of, 31–32, 35
  preferential differences in, 27–35, 41–42, 45
  signaling theory of, 43–44
  supertasters and, 28–29
foot fetishists, 67–68
Foreman, George, 149
Fore people, 37
forgeries and fakes, 2, 108, 119, 120–22, 140, 148, 149, 206, 210
Fountain (Duchamp), 143, 144
Fountain/After Marcel Duchamp (Levine), 146
4'33" (Cage), 144
Frank, Robert, 209
freak shows, 153
Free Willy 2, 188
Freud, Sigmund, 65, 71, 192, 196
Freud Museum in London, 101
Friday the 13th, 191, 195

Friends (TV show), 156, 160, 175
frozen yogurt, 46–47
fruit flies, 83
Furukawa, Stacy, 118
future planning, 161–63

Gacy, John Wayne, 105
Gajdusek, Carleton, 36–37
Galileo Galilei, 219–20
Gandhi, Mohandas, 41, 218
Garbo, Greta, 121
Gates, Bill, 140
Gelman, Susan, xiv, xv, xvi, 13, 14, 16, 17, 143
gender differences, 15, 71–73, 85
  biological essentialism and evolutionary origins of, 59–65, 72
  category essentialism and, 14, 15, 17, 59, 71–73
  children's perception of, 14, 17, 71–73
  cultural theories of, 72
  in human sexuality, 62–65, 66–67
  sex-role transgressions and, 72–73
  theory of parental investment in, 60–62, 63
Gendler, Tamar, xiv–xv, 169
General Mills, 42
Genesis, 77
genital mutilation, 79–80
gift cards, 95–96
Gil-White, Francisco, 13
Gladwell, Malcolm, 150–51
God Is Not Great (Hitchens), 215
Goering, Hermann, 1–2, 120–21
Goodstein, Joshua, xv, 188
Gopnik, Adam, 112
Gopnik, Alison, 214, 218
Gopnik, Olivia, 112
gorillas, 36
Gould, Stephen Jay, xiii, 11
Grand Canyon Skywalk, 169
Grand Theft Auto (video game), 190
Greeks, 148
Greenwich Museum, 205
groupthink, 119
Guinness Book of World Records, 152
gurning competitions, 153–54
Gyatso, Tenzin, 21–22

Haidt, Jonathan, 73, 177, 216, 218
Hamlet (Shakespeare), 191
Hamlin, Kiley, xv, 195–96
Harris, Judith, 32
Harris, Marvin, 30, 33
Harris, Sam, 215
Harry Potter series, 165–66, 168, 196
Hatano, Giyoo, 18–19
Hebrew Bible, 56–57, 77, 164
Hensel, David, 147
heterozygosity, 66
Heyman, Gail, 17
Hindu philosophy, 20
Hindu texts, 164
Hirst, Damien, 153
Hitchens, Christopher, 215
Hitler, Adolf, 1, 104, 105, 120, 218
Hochberg, Julian, 131
Holy Grail, 168
Homer, 15, 56
homosexuality, 58, 59, 62, 72
Hood, Bruce, xv, xvi, 104, 107–10, 111, 113–15, 205–6
horror movies, 156, 178, 181, 187, 191, 192, 194, 196
Hostel, 191
"hot chooser," 84–85
Hrdy, Sarah, 36
human pleasures, xi–xiv
  ability to disassociate choices from, 50

human pleasures (*continued*)
accidental forms of, xiii, 59, 124, 173, 218
in art and performance, *see* artwork; performance valuation
in awed emotional responses, 212, 216–18
as by-product of essentialist mindset, 8–9, 22, 49–50
in contacting the transcendent, xiv, 23, 212–14, 215, 216
cultural and social theories of, 6–8, 29–32, 34–35, 59, 128–29
depth of, xii, xiii, xiv, 23–24, 53, 89, 122, 125, 210
in everyday objects, *see* object valuation
evolutionary theories of function of, 4–8
in imagination, *see* imagination and imaginative pleasures
immoral types of, 209
inverted U rule of, 127, 129
masochism and, xi, 51–52, 156, 194–97, 199, 200, 208, 209
"mere exposure" effect in, 69–70, 127, 132–33
in music and dance, 5, 122–29
in nature and purity, 42–43, 68, 210–11
in pain, xi, 51–52, 178, 189, 190, 192, 194–97, 199, 200, 208, 209
pain from, 52–53
in performance, *see* performance valuation
in religion and spirituality, xi, xii, xiv, 212–14
in seeking out essences of things, 211–16, 218, 221
self-consciousness to, 50–51
in sex, *see* sexuality and sexual pleasures
signaling theory in, 43–44, 52, 83–84, 133
in tragic events, 174–75
unique to our species, xi, 5, 6, 122–23, 194
in virtuoso displays, 112, 137–38, 140–41, 142, 146, 149–50, 153, 154, 179–83, 209
in visual images, 58, 88, 119–22, 129–32, 133, 219; *see also* artwork; paintings
Hume, David, 113, 168, 190, 191
humor, 180–81, 182
Hungry Soul, The (Kass), 52
Hussein, Saddam, 105
hymen reattachment, 79

I Am Not Spock (Nimoy), 168
I Am Spock (Nimoy), 168
ice cream, 45, 46–47
"IKEA effect," 142
imaginary friends, xi, 198–99
imagination and imaginative pleasures, xi, xii, 155–94, 197–202, 218–21
alief experienced in, 169–71, 184, 186–88
appreciating virtuosity of creators in, 179–83
biological and evolutionary theories of, xiii–xiv, 158, 161–63, 171–73, 201
capacity to reason about another's false belief in, 160, 171, 218–19
catharsis in, 192
in children, xi, 156–57, 158–59, 163, 168, 185–88, 192–93, 198, 220, 221
confounding of reality with, 167–68
in daydreaming, 155, 156, 163, 170, 178, 197–201

distinguishing reality from, 156, 157–58, 163, 167, 168, 185–86, 187
emotional triggering by, xi, 165–71, 174–75, 177, 184, 186–88, 190–92
evolving tastes in, 182–83
in fiction and stories, xi, xii, xiii–xiv, 7, 124, 156, 163–68, 170, 171–76, 178–92, 193–94, 219, 220
future planning possible with, 161–63, 201, 218
metarepresentation and, 159–63
pain and fears explored in, xi, 178, 189, 190, 191–92, 193–95, 199, 200
power to be transported by, 170–71, 181
as practice for dealing with real-life horrors, 173, 193–94, 201
in pretend play, xiv, 156–58, 163, 170, 192–93, 220
psychic intimacy in, 175–76, 184
religious role of, 219, 220–21
safety in, 173, 183–85, 188–90, 193, 201
in scientific inquiry, 219–20, 221
as social learning tools, 171–73, 201
technology and increasing realism to, 178–79, 200–201
in video games, 155, 156, 170–71, 189–90, 200–201
of violent and sadistic scenarios, 181, 189–92, 193–94
Inagaki, Kayoko, 18–19
"inbreeding depression," 74
incest avoidance, 73–77
infanticide, 36
insects, 30, 33, 34, 35
International Flavors & Fragrances, 42
Introduction to Shakespeare (Johnson), 174
inverted U rule of pleasure, 127, 129
Isabella, Queen of Spain, 36
Ivory Coast, 131

Jacobs, A. J., 182
Jagger, Mick, 135
James, Clive, 175
James, William, 23, 63, 137, 165, 213
Jarudi, Izzat, xv, 150
Jaws, 124, 171
Johnson, Marcia, xv
Johnson, Samuel, 174
Johnson, Steven, 182–83
John the Baptist, 30
Jordan, Michael, 109, 153
Judaism, 13, 14, 39, 57, 220

Kahn, Peter H., Jr., 211
Kahneman, Daniel, 206
Kass, Leon, 20, 52
Keats, John, 103
Keil, Frank, xvi, 16
Kelly, Ellsworth, 118
Keltner, Dacher, 216–18
Kennedy, John F., 3, 19, 100, 101, 102, 115, 207
Kienholz, Ed, 153
King, Stephen, 193
Kleven, Deborah, 141
"kluges," 206
Koestler, Arthur, 119–20, 205, 206
Ko Samet, 162
kosher law, 38
Kruger, Justin, 141
kuru, 37

lactose intolerance, 28
language, 124, 125, 151, 164, 165, 171
essentialism as pervasive in, 10–12, 16–17
Latin, 43

laugh tracks, 182
Layard, Richard, 209
Lee, Leonard, 47
Leithauser, Mark, 118
Leonard, Elmore, 178
Leslie, Alan, 163
Levine, Sherrie, 146
Leviticus, 74
Levitin, Daniel, xvi, 125
Lewontin, Richard, xiii
Lieberman, Debra, 75
life-force essentialism, 18–22
  cannibalism and, 20, 26–27, 36–41, 206–7
  category vs., 20–21
  in everyday objects, 18–19, 102–5, 205, 214
  in religious rituals, 19, 21–22, 39, 214
  special contact and positive contagions in, 19–20, 102–5, 115, 140, 145
Lin Yutang, 170
Locke, John, 9
locusts, 30
London, 162
Lone Star, 74
Louisiana Museum of Modern Art, 153
Louvre Museum, 151

McEwan, Ian, 164–65
McFarlane, Todd, 3
McGinn, Colin, 176, 178
McGwire, Mark, 3, 208
Malraux, André, 140
mandrills, 85
Manson, Charles, 105
Manzoni, Piero, 145–47, 153
Mar, Raymond, 173
Marcus, Gary, 206
Marcus Welby, M.D., 168
Market Pricing, 94–97
Markson, Lori, 139
marmosets, 123
masochism, xi, 51–52, 194–97, 208, 209
  in daydreams, 156, 199, 200
Masochist Cookbook, 51
masturbation, 59, 78, 79, 170, 202
Matrix Corporation, 35
Matson, Katinka, xv
medial orbitofrontal cortex, 48
Meiwes, Armin, 25–26, 27, 53
Mekranoti people, 122–23
Memento, 11
Menand, Louis, 7, 146
menstrual cycles, 66–67
"mere exposure" effect, 69–70, 127, 132–33
metarepresentation, 159–63
Microsoft Flight Simulator (video game), 189
Middle Ages, 100
military, U.S., 35
Miller, Geoffrey, xv, xvi, 83–85, 134, 135, 136, 209
Miller, William Ian, 34
Milli Vanilli, 149
modern art, 43, 120, 133, 140–41, 146, 153
Molyneux, Juan, 207
money taboos, 93–97
monkeys, 123, 130, 170, 177
monogamy, 63, 64
Montaigne, 168–69
morality, 53, 92
movies, xi, 155, 163, 166, 176, 177, 178, 181, 184, 185
  horror, 156, 178, 181, 187, 191, 192, 194, 196

music, xi, xiv, 8, 49, 58, 122–29, 144, 149, 154, 171, 183
bonding and solidarity established with, 125–26
context in appreciation of, 117–18, 119
as "Darwinian fitness test," 135
emotional triggering by, 124
evolutionary and adaptationist theories of pleasure in, 122–26, 135
individual taste preferences in, 28, 126–29
inverted U rule of pleasure in, 127
"mere exposure" effect in, 127
as uniquely human pleasure, 5, 122–23
muti, 40–41
My Bed (Emin), 143

Napoléon, 101, 103, 115
National Gallery, 118
natural foods, 42, 44, 49, 53
natural selection, 4, 6, 7, 58, 59, 60, 82, 105, 206
see also evolutionary theories
Nature, 75
Nazis, 2
Nevins, Bruce, 44
New Age movement, 40
New Guinea, 134
Newman, George, xv, 103
New Testament, 77
New York City Marathon, 150
New Yorker, 160–61
New York Times, 191–92
Night Watch, The (Rembrandt), 119
Nimoy, Leonard, 168
NME, 37
Norton, Michael, 142
Nozick, Robert, 179
"numinous," 215Nureyev, Rudolf, 153
Nussbaum, Martha, 173
Nuttall, A. D., 161

Oates, Joyce Carol, 101
Oatley, Keith, 173
Obama, Barack, 14, 100, 103, 105
obesity, 203
object valuation, xii, 9, 87, 91–115
ability to reason about individuals in, 105–7
anthropomorphization in, 113
of celebrity memorabilia, xii, 3, 19, 99–101, 102, 103–5, 108, 110, 112, 115, 207
by children, xi, 87, 93, 99, 105–10, 113–15, 206, 208, 220–21
endowment effect in, 98–99
history and origins in, 3–4, 93, 99–105, 108–10, 115, 205, 207, 208
immoral side of, 209
life-force essentialism in, 18–19, 102–5, 115, 140, 145, 205, 214
money taboos and trade-offs in, 91–97
negative contact and, 104–5
perceived irrationality of, 205–9
personal history in, 97–99, 102
religious ritual and, 19, 21–22, 39, 214
sentimental attachment in, 3, 5, 87, 96, 102, 115, 206
special contact and positive contagions in, 19–20, 99–101, 102–5, 108, 110, 112, 115, 133, 140, 145, 207
"transitional" or attachment objects in, xi, xii, 113–15, 206, 208, 220–21
see also artwork; paintings
Odyssey (Homer), 56

Olmstead, Marla, 142
Omnivore's Dilemma, The (Pollan), 42
One (Number 31, 1950) (Pollock), 140–41
One Day Closer to Paradise (Hensel), 147
opiates, 52
Opie, Peter and Iona, 214
optimal foraging theory, 30
oral sex, 78
orange juice, 45
organ transplantation, 20
orgasm, female, 58

paintings, xii, 1–3, 7, 22, 132–34, 182, 183
  animal creation of, 144
  context in appreciation of, 118, 119
  creative intention in appreciation of, 143–45
  "mere exposure" effect in appreciation of, 132–33
  origins and history in appreciation of, 2–3, 87, 119–22, 138–47
  see also artwork
Papua New Guinea, 37
"paradox of horror," 191
parental investment, theory of, 60–62, 63
Parker, Dorothy, 63
Passover, 220
pathetic fallacy, 172
peacocks, 82–83, 134, 135
Pepsi, 45, 48
performance valuation, xii, 117–54
  biological essentialism and, 134–35, 136–37
  cheating and, 149–51
  competitiveness and, 152–54
  context in appreciation of, 117–19, 122
  creative intention in, 138–39, 143–45, 146, 210
  "Darwinian fitness displays" theory and, 133–37, 152–53, 154
  inverted-U rule of pleasure in, 127, 129
  "mere exposure" effect in, 127, 132–33
  originality and uniqueness as factor in, 148–49
  origins and history as factor in, 2–3, 119–22, 138–47, 154
  perceived effort in, 141–42
  and pleasure in virtuoso displays, 112, 137–38, 140–41, 142, 146, 149–50, 153, 154, 179, 183
  snobbery in, 119, 120
  in sports, 148–51, 152, 154
  theory for defining art in, 142–47
  in ugly and paradoxical displays, 153–54
  visual pleasure and, 129–31, 132
  see also artwork; music; paintings
Perrier, 44, 49, 50
pheromones, 69
physical attraction, 65–70, 82, 86, 129, 151
  familiarity in, 69–70
physiology, in food preferences, 28–29, 49
Piaget, Jean, 14–15
Picasso, Pablo, 119, 134, 136
pygmy chimpanzees, 164
Pinker, Steven, xv, 5, 86–87, 124, 173
Pitt, Brad, 172
Pittsburgh, University of, 69
placentas, eating of, 40
plastic surgery, 68, 151, 210
Plato, 194
Playboy, 69

play-fighting, 158, 192–93
Pollan, Michael, 42
Pollock, Jackson, 130, 140–41, 146
polygamy, 63, 64
Popper, Karl, 161
pornography, xiii, 6, 62, 88, 130, 156, 191
positive contagions, 19–20, 102–4, 110, 112, 115
  in artwork valuation, 133, 140, 145
postmodern art, 146
postpartum depression, 40
Powers, Richard, 101
prenuptial agreements, 84
pretend play, xiv, 156–58, 163, 170, 192–93, 220
Procter & Gamble, 207
PROP test, 28, 29
prostitution, 62
protein bars, 45
psychic intimacy, in fiction, 175–76

Rabid, 191
race, 13, 15
Rape Lay (video game), 190
rats, 5, 30, 31, 123, 159
realistic pictures, 129–32
reality television, 7, 176
Real World, The (TV show), 176
Rear Window, 176
red wine, 46
religion, xi, xii, xiv, 5, 19, 21–22, 126, 133, 215, 219
  cannibalism in, 39
  contacting the transcendent in, 23, 212–14
  female virginity in, 77–78
  imagination's role in, 219, 220–21
Rembrandt van Rijn, 119, 120, 146
Renoir, Pierre-Auguste, 134
Republic, The (Plato), 194
rhesus monkeys, 130
Richards, Keith, 37
Rijksmuseum, 119
rituals, 19, 38, 212–14, 220
  cannibalism in, 38, 39
  in dining etiquette, 52
road rage, 204
Robinson, Ken, 221
Rocker, John, 11
Rodin, Auguste, 144
Rolling Stone, 151
Rolling Stones, 4
Romanes, George, 5
romantic love, 59, 86–88
Rose, Carol, xv
Rowling, J. K., 165–66, 168
Royal Academy of Arts (London), 147
Rozin, Paul, xiv, xv, xvi, 28, 33, 34, 41, 51, 52, 104, 131, 169, 194, 195
Rubin, Gretchen, 20
Ruiz, Rosie, 150
Ruskin, John, 172

sadomasochism, 196, 197
Salinger, J. D., 8–9
Sam Adams, 47
Santos, Laurie, xv, xvi, 99
Sapolsky, Robert, 128
Saw series, 191
Sayles, John, 74
schizophrenia, 198
science, xii, 58, 213
  essences uncovered in, 10, 12, 15, 211–12, 214, 215
  imagination used in, 219–20, 221
  transcendence and, 23, 214, 215
Scotland, 168
Second Life, 200, 201
security blankets, xi, xii, 5, 206, 208

self-mutilation, xi, 195
self-perception theory, 99
self-punishment, 195–96
Sendak, Maurice, 39
sentimental objects, 3, 5, 87, 96, 102, 115, 206
sex drive, 58, 59
sexual desire, 88–90
sexuality and sexual pleasures, xi–xii, 5, 6, 7, 22, 55–90, 155, 208, 209
  animal reproductive habits and, 57–61, 62, 67, 82–83, 85, 134, 135
  bedtricks and, 55–57, 88, 164
  biological essentialism in, 57–67, 74, 76, 77, 83, 134–35
  category essentialism in, 71–80, 206
  defining "sexual relations" in, 78–79
  essentialist nature of romantic and sexual desire in, 57, 68, 86–90, 111
  evolutionary theories of, 57–65, 69, 74, 77, 79, 82–85
  female virginity's importance in, xi, 77–80, 206, 209
  fetishes in, 58, 67–68
  gender differences in, 59–65, 66–67, 85
  incest avoidance in, 73–77
  masochism in, xi, 196, 197, 208
  "mere exposure" effect in, 69–70, 127
  personal history and, 59
  physical attraction in, 65–70, 82, 86, 129, 151
  in pornography, xiii, 6, 62, 88, 130, 156
  romantic love in, 59, 86–88
  sexual selection theory of, 63–64, 82–85, 86, 133–36
  theory of parental investment in, 60–62, 63
sexual selection theory, 63–64, 82–85, 86
  of origins of art and music, 133–36
sexual snobs, 120
Shakespeare, William, 55, 90, 100–101, 174, 199
Shakur, Tupac, 100
sham nudity, 130
sham virginity, 79
Sharkey, Lorelei, 78, 79
Shaw, George Bernard, 86
Shelley, Percy Bysshe, 103
shellfish, 33
sibling incest, 73–76
signaling theory, 52, 133, 140
  in food preferences, 43–44
  in sexual selection, 83–84
Simpson, O. J., 174, 175
Sims, The (video game), 189–90
Singer, Isaac Bashevis, 88, 89, 101
Singer, Peter, 209
60 Minutes II, 142
Slate, 78
smiling, 70
Smith, Ewart E., 35
Smith, John Maynard, 83
Smith, Susan, 174
Smith, Zadie, 101
snobbery, 119, 120, 122
Sobel, David, xv, 188
social adaptations, 125–26, 161–63, 217–18
socialization theories, 17
  evolutionary vs., 6–8
  of food preferences, 29–32, 34–35
  of gender differences, 72
  and incest avoidance, 74–77
  of musical taste preferences, 128–29
  see also cultural theories
social snobs, 120

Sontag, Susan, 101
Sophie's Choice (Styron), 136–37
Sopranos, The (TV show), 164
Sotheby's, 121
Spam, 32
spandrels, xiii
Spears, Britney, 100
"sperm wars," 64
Spielberg, Steven, 199
spirituality, 213, 215, 216
sports, xiv, 122, 148–51, 152, 154, 183, 193
Sports Illustrated, 151
Stalin, Joseph, 218
standup comedy, 152
Star Trek: The Next Generation, 100
stepfathers, 76–77
steroids, 150–51
"stigmatized homes," 104
stories, *see* fiction
Stradivari, Antonio, 117
Styron, William, 136–37
sublime reactions, 216
SuperSense (Hood), 104
supertasters, 28–29
Supper at Emmaus, The (van Meegeren), 2, 120, 121, 122
Supreme Court of Judicature, 207
Survivor (TV show), 176
Sweet Hereafter, The, 194
symmetry, in physical attractiveness, 65, 66, 68–69, 70, 82
Symphony Hall (Boston), 117

"taboo trade-offs," 91–92, 96–97
Taiwan, 75
Tajfel, Henri, 12
tamarins, 123
Tanzania, 40–41
taste, 44–50
Tate Gallery, 143, 145, 146
Taylor, Emma, 78, 79
Taylor, Marjorie, xvi, 71–72, 198–99
teledildonics, 88–89
television, xi, 7, 40, 155, 156, 164, 167, 168, 175, 176, 203
  evolving programming on, 182–83
Terms of Endearment, 166, 194
Tetlock, Philip, 92
Thailand, 162
theism, 4
Thomas, Sonya "The Black Widow," 153
"thought experiments," 219–20
toilet training, 34
tongue physiology, 29
Tooby, John, 75
torture porn, 191
tragedy, pleasure in, 174–75, 190, 191, 192
Trainspotting, 166
transactions, 91–97
transcendent pleasure, xiv, 23, 212–15, 216, 221
transexuals, 72
"transitional objects," 113, 220–21
Trivers, Robert, 60
Turgenev, Ivan, 172
turkeys, 67
TV Dinners (British series), 40
Tversky, Amos, 206
24 (TV show), 182–83
Twilight Zone (TV show), 200
Tylor, Edward Burnett, 212

ugly art, 153
universal plots, 7, 164–65
Updike, John, 101

van Meegeren, Han, 1–2, 120–21, 142
Varieties of Religious Experience (James), 213

vasectomies, 84
Veblen, Thorstein, 43
vegetarianism, 41, 42
Vermeer, Johannes, 1, 2, 120–21, 142
vervets, 85
Viagra, 41
video games, 7, 153, 155, 156, 170–71, 189–90, 200–201
virginity, xi, 77–80, 206, 209
virtual worlds, 200–201
visual pleasures, 58, 119–22, 129–32, 133, 219
  experimental aesthetics of, 129
  in pornography, xiii, 88, 130
  realistic images preferred in, 129–31
  see also artwork; paintings
von der Lippe, Angela, xv, xvi
Vonnegut, Kurt, 123

Wallace, David Foster, 101
Walzer, Michael, 91–92
Warhol, Andy, 144, 145
Washington, D.C. snipers, 174
Washington Post, 117
water, bottled, 42–43, 44, 49, 50, 83, 210
Wedgwood, Emma, 80–82, 163
Weinberg, Steven, 213
Weingarten, Gene, 117–18
Weisberg, Deena Skolnick, xv, xvi, 186, 188
West, Fred and Rosemary, 104
Westermarck, Edward, 75
"What Money Can't Buy" (Walzer), 91–92
"Where the Hell Is Matt?" (YouTube video), 126
Where the Wild Things Are (Sendak), 39
white wine, 46
Why Is Sex Fun? (Diamond), 62
Wilson, E. O., 204
wine, 29, 45–46, 48
Winger, Debra, 166
Winnicott, Donald, 113, 220–21
wolves, 158
World of Warcraft, 200
World Without Us, The (Weisman), 171
Wright, Richard, 178
Wynn, Karen, xv, xvi, 106–7, 195–96

Yale University, xv, 145
Yankee Stadium, 151
Yenawine, Philip, 140–41
yogurt, 45
Young, Robert, 168
Young Christ Teaching in the Temple, The (van Meegeren), 2
Young Woman Seated at a Virginal (Vermeer), 121
YouTube, 126

Zodiac, 167
Zunshine, Lisa, xvi, 160, 172, 173